Umurdin Dalabaev

Solution of Initial-Boundary Value Problems

De Gruyter Series in Applied and Numerical Mathematics

Volume 11

Umurdin Dalabaev

Solution of Initial-Boundary Value Problems

Method of Moving Modes

DE GRUYTER

Mathematics Subject Classification 2020
Primary: 65N06; Secondary: 65L10

Author
Prof. Umurdin Dalabaev
Department of System analysis and mathematical modeling
University of World Economy and Diplomacy
st. Buyuk Ipak yuli, 54
Tashkent 100007
Uzbekistan
udalabaev@mail.ru

ISBN 978-3-11-130439-7
e-ISBN (PDF) 978-3-11-130817-3
e-ISBN (EPUB) 978-3-11-130873-9
ISSN 2512-1820

Library of Congress Control Number: 2025931480

Bibliographic information published by the Deutsche Nationalbibliothek
The Deutsche Nationalbibliothek lists this publication in the Deutsche Nationalbibliografie; detailed bibliographic data are available on the internet at http://dnb.dnb.de.

Typesetting: Integra Software Services Pvt. Ltd.
Printing and binding: CPI books GmbH, Leck

www.degruyter.com
Questions about General Product Safety Regulation:
productsafety@degruyterbrill.com

Contents

Introduction

Mathematical modeling of many problems in mechanics, physics, chemistry, and other fields of science and technology leads to differential equations, ordinary or partial. When solving problems of mathematical modeling, it is often necessary to be able to solve these equations.

Methods for solving problems of mathematical physics can be divided into the following four classes:

Analytical methods (method of separation of variables, method of characteristics, method of Green's functions, etc.) have a relatively low degree of universality, that is, they are aimed at solving rather narrow classes of problems. As a result of applying these methods, a solution is obtained in the form of analytical formulas. Using these formulas to perform a calculation may require solving auxiliary computational problems (solving nonlinear equations, calculating special functions, numerical integration, summation of an infinite series). Nevertheless, in some cases, the use of these methods allows one to quickly and accurately calculate the desired solution.

Approximate analytical methods (projection, variational methods, small parameter method, operational methods, and various iterative methods) are more universal than analytical methods. The use of such methods involves modifying the original problem or changing the formulation of the problem so that the new problem can be solved by an analytical method, and its solution itself differs little from the solution of the original problem. Numerical methods (finite-difference method, method of lines, control volume method, finite element method, etc.) are very universal methods. They are often used to solve nonlinear problems of mathematical physics as well as linear problems with variable operator coefficients.

Probabilistic methods (Monte Carlo methods) are highly versatile. They can be used to calculate discontinuous solutions. However, they require large amounts of computation and, as a rule, are inferior in computational complexity to the above methods when solving problems to which these methods are applicable.

When comparing methods for solving problems in mathematical physics, it is impossible to give absolute priority to any of them. Any of them may be the best for solving problems of a certain class. At the same time, when characterizing a specific method, it is advisable to highlight those of its features that often determine its advantages or disadvantages in practical application compared to an alternative method.

The advantages of the finite-difference method include its high versatility, for example, significantly higher than that of analytical methods. The use of this method is often characterized by the relative simplicity of constructing a solving algorithm and its software implementation. It is often possible to parallelize the solving algorithm.

The disadvantages of the method include the problematic nature of its use on irregular grids; very rapid growth of computational complexity with increasing problem

https://doi.org/10.1515/9783111308173-001

dimension (increasing number of unknown variables); complexity of analytical study of difference scheme properties.

The proposed method of moving nodes combines numerical and analytical methods. In this case, we can obtain, on the one hand, an approximately analytical solution to the problem, which is not related to the methods listed above. On the other hand, this method allows us to obtain compact discrete approximations of the original problem. Note that obtaining an approximately analytical solution to differential equations is based on numerical methods. The nature of numerical methods also allows obtaining an approximate analytical expression for solving differential equations. For this purpose, the so-called movable node is introduced.

The purpose of the study is to develop a computational technology for solving differential equations based on the proposed method of moving nodes and show the capabilities of the method for solving boundary and initial-boundary value problems of ordinary and partial differential equations.

The first chapter presents the necessary known facts related to the topic under study.

The second chapter reflects the results of research applied to two-point boundary value problems.

The applications of the method of moving nodes of the Cauchy problem for ordinary differential equations are given in the third chapter.

The fourth chapter is devoted to obtaining an approximately analytical method for parabolic equations based on moving nodes.

The results of applying the method of moving nodes for elliptic equations are given in the fifth chapter.

Chapter 1
Basics of numerical methods

This chapter presents some facts known in the literature. The necessary materials on which the subject of this study is based are presented. The concept of a grid, grid functions, and finite-difference methods (FDMs), as well as the control volume method, is briefly presented. There are sufficient numbers of publications devoted to numerical methods for differential equations [1, 22, 25, 27, 28].

1.1 Grids and grid functions: introductory concepts

The grid method or difference method is one of the effective numerical methods for approximate solution of equations. In the grid method, the domain of continuous variation of arguments is replaced by a discrete set of points, which is called a grid. Grid functions are functions defined at the grid nodes. Derivatives appearing in differential equations and boundary conditions are replaced by difference relations. After discretization, the differential equation is replaced by a system of linear or nonlinear algebraic equations. The resulting systems are called difference schemes. Briefly, let us outline the basic concepts of the theory of difference schemes.

One-dimensional case: A grid on a segment $x \in [a, b]$ is any finite ordered set of points of this segment. Most often, a uniform grid of the form will be used:

$$\bar{\omega}_h = \{x_i = a + ih; \quad i = 0, N\},$$

where $h = (b - a)/N$ is the grid step.

Two-dimensional case: For a rectangle $G = \{a \leq x \leq b;\ c \leq y \leq d\}$ we define a uniform grid as $\bar{\omega}_h = \bar{\omega}_1 \times \bar{\omega}_2$, where

$$\bar{\omega}_1 = \{x_i = a + ih_1;\ i = 0, N_1\} \text{ and } \bar{\omega}_2 = \{y_j = c + jh_2;\ j = 0, N_2\}.$$

Here $h_1 = (b - a)/N_1$ and $h_2 = (b - a)/N_2$ are the steps in the variables x and y. If $h_1 = h_2$, then the grid is called square; otherwise – rectangular. In the general case, the grid may be nonuniform.

Let us consider the concept of a grid function. Let $\bar{\omega}_h$ be a grid introduced in the one-dimensional case and x_i be grid nodes. The function $y = y(x_i)$ of a discrete argument x_i is called a grid function defined on the grid $\bar{\omega}_h$. A grid function on any grid is defined similarly. Grid functions can be considered as functions of an integer argument, which is the number of a grid node.

Let a grid $\bar{\omega}_h$ be given. The set of all grid functions defined on the grid forms $\bar{\omega}_h$, a vector space. In the space of grid functions, one can define difference or grid operators.

https://doi.org/10.1515/9783111308173-002

The operator Λ that transforms a grid function y into a grid function $f = \Lambda y$ is called a difference or grid operator. The set of grid nodes used to write a difference operator is called the template of this operator. The simplest difference operator is the operator of difference differentiation of a grid function.

Let us consider the problem of approximately calculating the derivatives of a function $u(x)$ defined on a segment $[a, b]$. We will assume that the function $u(x)$ has a certain degree of smoothness, and then the first-order difference derivatives on the grid $\bar{\omega}_h$ for the function $u(x)$ are given in Tab. 1.1 on a uniform grid.

Tab. 1.1: Difference operators approximating the derivative.

Designations	Types of difference operators and the main error terms	Approximation order
Λ_-	$\Lambda_- u_j = \frac{u_j - u_{j-1}}{h} = (u_x)_j - \frac{h}{2}(u_{xx})_j + \frac{h^2}{6}(u_{xxx})_j + O(h^3)$	$O(h)$
Λ_+	$\Lambda_+ u_j = \frac{u_{j+} - u_j}{h} = (u_x)_j + \frac{h}{2}(u_{xx})_j + \frac{h^2}{6}(u_{xxx})_j + O(h^3)$	$O(h)$
Λ	$\Lambda u_j = \frac{u_{j+1} - u_{j-1}}{2h} = (u_x)_j + \frac{h}{6}(u_{xxx})_j + O(h^4)$	$O(h^2)$
Λ_{2-}	$\Lambda_{2-} u_j = \frac{3u_j - 4u_{j-1} + u_{j-2}}{2h} = (u_x)_j - \frac{h^2}{3}(u_{xxx})_j + O(h^3)$	$O(h^2)$
Λ_{2+}	$\Lambda_{2+} u_j = \frac{-3u_j - 4u_{j+1} + u_{j+2}}{2h} = (u_x)_j - \frac{h^2}{3}(u_{xxx})_j + O(h^3)$	$O(h^2)$

If the point x_i is fixed and the step h tends to zero, then each of the difference relations tends to the value of the derivative of the function $u(x)$ at the point x_i. Note that the derivative can be replaced by any difference relation, but the order of the error with such a replacement will be different.

Higher-order difference derivatives can be introduced using recurrent formulas. If in the formulas for the first-order difference derivative, instead of the values of the function at the grid node, we use the value of the difference derivative, and we obtain formulas for calculating the second-order difference derivative, for example:

$$u_{xx,i} = \frac{u_{x,i+1} - u_{x,i}}{h} = \frac{u_{i+1} - 2u_i + u_{i-1}}{h^2}$$

The second-order difference derivative is not unique. Difference derivatives of any order are constructed similarly.

Examples of the simplest approximations of differential expressions by difference ones on a uniform grid are given. In the general case, the error depends both on the distribution of grid nodes and on the smoothness of the function.

The following presentation is devoted to the construction and study of difference boundary value problems for model equations of mathematical physics.

1.2 Difference schemes for the boundary value problem of an ordinary differential equation

1.2.1 Finite-difference method

The FDM is one of the oldest numerical methods for obtaining approximate solutions of partial differential equations.

The FDM is a direct approximation of the resolving partial differential equations, which is performed by replacing partial derivatives with differences on regular or irregular grids defined on the problem domain. Thus, in this case, a transition is made from a system of partial differential equations to a system of algebraic equations with respect to unknown quantities at grid points. The solution to the system of equations is obtained after imposing the necessary initial and boundary conditions. Therefore, the FDM is also called the grid method.

The main principle of the FDM is to replace partial derivatives of an unknown function with finite differences. As a result of this procedure, a system of algebraic equations is obtained for unknown functions defined at the nodes of the grid into which the problem domain is divided. The solution of a system of algebraic equations, including boundary conditions specified at the boundary nodes, yields the desired values of the unknown function at all nodes that will satisfy both the resolving system of partial differential equations and the specified boundary conditions.

Thus, the fundamental characteristic of the FDM is that it uses direct discretization of the resolving differential equations by replacing partial derivatives with finite differences defined at adjacent grid nodes. Creating a finite-difference grid is only a convenient way to represent the values of the sought functions at specified points located at a sufficiently small distance from each other, such that the resulting errors in the solution are sufficiently insignificant and can be neglected.

In addition, the FDM does not use approximating or interpolating functions to represent differential equations in the vicinity of specified points, as is the case, for example, in the finite element method and the boundary element method.

The FDM thus represents the most direct and visual method for solving partial differential equations.

Consider a linear two-point boundary value problem.

Consider a two-point boundary value problem for a linear differential equation. Find a solution to the equation:

$$y'' + p(x)y' + q(x)y = f(x), \tag{1.1}$$

with additional boundary conditions

$$\begin{aligned} \alpha_0 y'(a) + \alpha_1 y(a) &= A, \\ \beta_0 y'(b) + \beta_1 y(b) &= B, \end{aligned} \tag{1.2}$$

where the numbers $\alpha_0, \alpha_1, \beta_0, \beta_1$ are considered known and

$$\alpha_0^2 + \alpha_1^2 \neq 0, \beta_0^2 + \beta_1^2 \neq 0;$$

that is, one of the quantities is not zero.

The coefficients $p(x), q(x), f(x)$ are continuous functions on a certain interval $[a, b]$. The solution to this equation is a function continuous on $[a, b]$, having first and second derivatives on $[a, b]$, satisfying the original equation and additional boundary conditions.

The posed boundary value problem is solved by passing from the original problem to a new one, written in finite-difference form. Then the solution to the new problem will be an approximate solution to the original problem. Since the first and second derivatives included in the equation and in the boundary conditions will be replaced by approximate finite-difference formulas, solutions using the FDM are obtained not in the form of a continuous function $y(x)$, but in the form of a table of its values at individual points. To do this, we divide $[a, b]$ into parts so that $a = x_0, b = x_n, x_{i+1} = x_i + h, h = (b - a)/n$. Our task is to find the values of the function $y(x)$ at the points x_k, $k = \overline{0, n}$. In order to move from the original problem to a finite-difference problem, we must obtain formulas for representing the first and second derivatives in finite-difference form. They are obtained by applying the expansion of the function $y(x)$ in the neighborhood $(\pm h)$ of a certain point $x_i \in [a, b]$ in a Taylor series, limiting ourselves to the second derivatives:

$$y(x_i + h) \cong y(x_i) + hy'(x_i) + \frac{h^2}{2} y''(x_i),$$

$$y(x_i - h) \cong y(x_i) - hy'(x_i) + \frac{h^2}{2} y''(x_i).$$

We add these series and obtain an expression for the second derivative in finite-difference form:

$$y(x_i + h) + y(x_i - h) \cong 2y(x_i) + h^2 y''(x_i),$$

$$y''(x_i) \cong \frac{y(x_i + h) - 2y(x_i) + y(x_i - h)}{h^2}.$$

Similarly, we obtain a formula for the first derivative if we subtract the series:

$$y(x_i + h) - y(x_i - h) \cong 2hy'(x_i),$$

$$y'(x_i) \cong \frac{y(x_i + h) - y(x_i - h)}{2h}.$$

Let us denote: $y(x_i) = y_i, y(x_i + h) = y(x_{i+1}) = y_{i+1}; y(x_i - h) = y(x_{i-1}) = y_{i-1}: p(x_i) = p_i, q(x_i) = q_i;$ $f(x_i) = f_i$.

Taking into account the introduced notations, we write the initial equation for the nodal points:

$$\frac{y_{i+1}-2y_i+y_{i-1}}{h^2}+p_i\frac{y_{i+1}-y_{i-1}}{2h}+q_iy_i=f_i \quad i=\overline{1,n-1}. \tag{1.3}$$

Represent

$$y'(a)=y'(x_0)=\frac{y(x_0+h)-y(x_0)}{h}=\frac{y_1-y_0}{h};$$

$$y'(b)=y'(x_n)=\frac{y(x_n)-y(x_n-h)}{h}=\frac{y_n-y_{n-1}}{h}.$$

In finite-difference form, two more equations corresponding to the boundary conditions are added to system:

$$\alpha_0\frac{y_1-y_0}{h}+\alpha_1y_0=A, \tag{1.4}$$

$$\beta_0\frac{y_n-y_{n-1}}{h}+\beta_1y_n=B. \tag{1.5}$$

We obtain a system of linear algebraic eqs. (1.3)–(1.5) with unknowns $y_0, \ldots, y_n$. Having solved this system by any known method, we obtain an approximate solution $y_0, \ldots, y_n$ for the original problem.

Note that the system is a system with a sparse matrix of tridiagonal form. Therefore, special methods are used to solve the system, which allow one to operate only with matrix elements that are different from zero. One of such methods is the Thomas algorithms [1, 12, 22].

1.2.2 The control volume method

The finite volume method or the control volume method, like the FDM, also directly approximates partial differential equations, but in the integral sense.

Various physical processes (heat or vibration propagation in a medium, gas dynamics, and aerodynamics processes) are characterized by some integral conservation laws (of heat, mass, momentum, and energy). When deriving differential equations of mathematical physics, one usually starts from some integral relation (balance equation) expressing the conservation law for a small volume. The differential equation is obtained from the balance equation when the volume is contracted to zero under the assumption of the existence of continuous derivatives included in the equation.

One of the main properties of a difference scheme is its conservatism. A difference scheme is called conservative if the difference analogs of physical conservation laws are satisfied for it. To obtain conservative difference schemes, it is natural to start from the balance equations written for elementary volumes (cells) of the grid domain. The integrals and derivatives included in these balance equations should be replaced by approximate difference expressions. As a result, we obtain a homogeneous difference scheme. This method of obtaining conservative schemes will be called the integro-interpolation method (balance method). Thus, the method allows us to construct difference schemes that satisfy the basic conservation laws if the corresponding continuous models satisfy these laws.

We will consider the method using the example of a boundary value problem for a second-order ODE of the form:

$$(k(x)u'(x))' - q(x)u(x) + f(x) = 0, \qquad 0 < x < l; \tag{1.6}$$

$$-k(0)u'(0) + \beta u(0) = \mu_1; \tag{1.7}$$

$$u(l) = \mu_2. \tag{1.8}$$

In eq. (1.6) and condition (1.7), the desired function is $u(x)$; $x \in [0, l]$. The functions $k(x)$, $q(x)$, $f(x)$ and constants β, μ_1, μ_2 are given. With respect to these functions, we will assume that

$$k(x) \geq C > 0,\ q(x) \geq 0, \quad x \in [0, l] \text{ and } \beta \geq 0.$$

In addition, we will assume that all functions have the smoothness required in the course of the presentation. Under the specified conditions, the solution to problems (1.6)–(1.8) exists and is unique.

Equation (1.6) can be considered as an equation for the steady-state distribution of temperature $u(x)$ in a rod of length l, at one end of which a given temperature is maintained, and at the other end heat exchange with the environment occurs.

To construct a solution to the difference scheme, we will use a uniform grid:

$$\omega_h = \{x_i = ih \quad i = \overline{1, N-1}, \quad Nh = l\};$$

$$\overline{\omega_h} = \omega_h \cup \{x = 0\} \cup \{x = l\};$$

$$\omega_h^+ = \omega_h \cup \{x = l\}; \quad \omega_h^- = \omega_h \cup \{x = 0\};$$

$$x_{i \pm 1/2} = x_i \pm 0,5h.$$

We integrate both parts of eq. (1.6) over the segment

$$x_{i-\frac{1}{2}} \leq x \leq x_{i+\frac{1}{2}}; \quad i = \overline{1, N-1}.$$

$$k(x)u'(x)\Big|_{x=x_{i+1/2}} - k(x)u'(x)\Big|_{x=x_{i-1/2}} - \int_{x_{i-\frac{1}{2}}}^{x_{i+\frac{1}{2}}} q(x)u(x)dx + \int_{x_{i-\frac{1}{2}}}^{x_{i+\frac{1}{2}}} f(x)dx = 0. \tag{1.9}$$

We will assume that the $u(x)$ is temperature. Heat flow inside the body is

$$Q = -k(x)\frac{du}{dx}.$$

Equation (1.9) is a slightly modified form of heat exchange according to Newton's law.

We introduce the heat flow function:

$$W(x) = k(x)\frac{du}{dx}. \tag{1.10}$$

Then eq. (1.9) can be rewritten as

$$W_{i+1/2} - W_{i-1/2} - \int_{x_{i-\frac{1}{2}}}^{x_{i+\frac{1}{2}}} q(x)u(x)dx + \int_{x_{i-\frac{1}{2}}}^{x_{i+\frac{1}{2}}} f(x)dx = 0. \tag{1.11}$$

We use approximate substitutions (1.12)–(1.14):

$$\int_{x_{i-\frac{1}{2}}}^{x_{i+\frac{1}{2}}} q(x)u(x)dx \simeq u_i \int_{x_{i-\frac{1}{2}}}^{x_{i+\frac{1}{2}}} q(x)dx \tag{1.12}$$

$$\varphi_i = \frac{1}{h}\int_{x_{i-\frac{1}{2}}}^{x_{i+\frac{1}{2}}} f(x)dx \tag{1.13}$$

$$d_i = \frac{1}{h}\int_{x_{i-\frac{1}{2}}}^{x_{i+\frac{1}{2}}} q(x)dx. \tag{1.14}$$

We obtain from eq. (1.11) an approximate equality, which we replace with an exact one, without changing the previous notations for simplicity:

$$W_{i+1/2} - W_{i-1/2} - hd_i u_i + h\varphi_i = 0, \qquad 1 \le i \le N-1. \tag{1.15}$$

We express the value of the function $W_{i\pm1/2}$ using the values of the function $u(x)$ at the grid nodes. From eq. (1.10) we obtain that

$$\frac{du}{dx} = \frac{W(x)}{k(x)}.$$

Integrating the relation on the segment $[x_i, x_{i+1}]$, we have

$$u_{i+1} - u_i = \int_{x_i}^{x_{i+1}} \frac{du}{dx}dx = \int_{x_i}^{x_{i+1}} \frac{W(x)}{k(x)}dx \approx W_{i+1/2} \cdot \int_{x_i}^{x_{i+1}} \frac{1}{k(x)}dx \overset{(10)}{=} W_{i+1/2} \cdot \frac{h}{a_{i+1}}.$$

Here

$$a_{i+1} = \left(\frac{1}{h} \int\limits_{x_i}^{x_{i+1}} \frac{dx}{k(x)} \right)^{-1}. \tag{1.16}$$

Proceeding to exact equalities, we obtain

$$u_{i+1} - u_i = W_{i+1/2} \cdot \frac{h}{a_{i+1}}. \tag{1.17}$$

Similarly

$$u_i - u_{i-1} = W_{i-1/2} \cdot \frac{h}{a_i}. \tag{1.18}$$

Substituting (1.17) and (1.18) into equality (1.15) yields:

$$a_{i+1}\frac{u_{i+1} - u_i}{h} - a_i \frac{u_i - u_{i-1}}{h} - hd_i u_i + h\varphi_i = 0 \quad 1 \le i \le N-1.$$

Divide both parts of the resulting equality by h:

$$\frac{1}{h}\left[a_{i+1}\frac{u_{i+1} - u_i}{h} - a_i \frac{u_i - u_{i-1}}{h} \right] - d_i u_i + \varphi_i = 0 \quad 1 \le i \le N-1. \tag{1.19}$$

The difference approximation of eq. (1.6) is constructed.

Compared to the direct approximation, some peculiarities arose in calculating the coefficients of the difference equations:

$$a_i = \left(\frac{1}{h} \int\limits_{x_{i-1}}^{x_i} \frac{dx}{k(x)} \right)^{-1}, d_i = \frac{1}{h} \int\limits_{x_{i-\frac{1}{2}}}^{x_{i+\frac{1}{2}}} q(x)dx, \quad \varphi_i = \int\limits_{x_{i-\frac{1}{2}}}^{x_{i+\frac{1}{2}}} f(x)dx.$$

Approximation of the boundary condition (1.8) is obvious:

$$u_N = \mu_2. \tag{1.20}$$

With the boundary condition (1.7), it is necessary to approximate. To obtain its discrete analog, we use the control volume method [1, 11].

Integrate both parts of eq. (1.6) over the segment $[o, h/2]$, then

$$W_{1/2} - W_0 - \int\limits_0^{h/2} q(x)u(x)dx + \int\limits_0^{h/2} f(x)dx = 0.$$

From condition (1.7): $W(0) = k(0)u'(0) = \beta u(0) - \mu_1 = \beta u_0 - \mu_1$.

From condition (1.17): $W_{1/2} = a_1 \frac{u_1 - u_0}{h}$. Substituting the values into the previous equation, we obtain

$$a_1 \frac{u_1 - u_0}{h} - \beta u_0 + \mu_1 - \frac{h}{2} u_0 d_0 + \frac{h}{2} \varphi_0 = 0, \tag{1.21}$$

where

$$d_0 = \frac{2}{h} \int_0^{h/2} q(x)dx; \ \varphi_0 = \frac{2}{h} \int_0^{h/2} f(x)dx.$$

Formula (1.21) approximates the boundary condition (1.7). Thus, the difference schemes (1.19), (1.20), and (1.21) are constructed.

The resulting difference boundary condition (1.21) can be written as

$$-a_1 u_{x,0} + (\beta + 0,5hd_0)u_0 = \mu_1 + 0,5h\varphi_0. \tag{1.22}$$

Thus, the boundary condition (1.7) has a difference analog equation (1.22).

In what follows, the solution of the difference problem, in contrast to the solution of the differential problem, will be denoted by the letter y so that $y_i = y(x_i)$, $x \in \omega_h$.

Combining eqs. (1.19) and (1.22), we obtain a discrete model for problems (1.6)–(1.8):

$$\frac{1}{h}\left[a_{i+1}\frac{u_{i+1} - u_i}{h} - a_i \frac{u_i - u_{i-1}}{h}\right] - d_i u_i + \varphi_i = 0 \qquad 1 \le i \le N-1; \tag{1.23}$$

$$-a_1 u_{x,0} + (\beta + 0{,}5hd_0)u_0 = \mu_1 + 0{,}5h\varphi_0; \qquad y_N = \mu_2.$$

1.3 Numerical solution of the Cauchy problem for an ordinary differential equation

All methods for solving the Cauchy problem for an ODE are divided into exact, approximate, and numerical. Exact methods are studied in courses on differential equations and allow one to express the solution of an equation through elementary functions or using quadratures of elementary functions. The class of problems whose solution can be obtained using exact methods is relatively narrow.

When using approximate methods, the solution of the Cauchy problem for an ODE is defined as the limit of a certain sequence of functions. In this case, each member of the sequence is expressed through elementary functions or quadratures of elementary functions. Approximate methods include expansion of the solution into a generalized power series, the Chaplygin method, the Picard method, and Kantorovich [25]. Approximate methods are convenient to use when it is possible to find an explicit expression for the coefficients of the series.

1.3.1 Explicit Euler method

Consider the Cauchy problem for ordinary differential equations (ODEs).

The simplest ODE has the form

$$y' = f(x, y(x)). \tag{1.24}$$

For it, the Cauchy problem can be posed: find a solution $y = y(x)$, $x \in [a,b]$, satisfying (1.24) and the initial condition

$$y(a) = y_0. \tag{1.25}$$

The existence and uniqueness of a solution to problems (1.24)–(1.25) follow from the Cauchy theorem [27].

Numerical methods are applicable to wide classes and systems of equations. With the advent of computers, these methods have become one of the main methods for solving practical problems for ODE. They do not provide a general solution to the problems and are intended to calculate approximate (and sometimes exact) values of particular solutions to problems (1.24) and (1.25) on a preselected grid of argument x – a finite set of points $x \in [a, b]$. As a result, the desired solution is obtained in the form of a table.

Let us describe the simplest of the numerical methods – the Euler method for finding a table of values of the solution to the Cauchy problem (1.24) and (1.25) for $x \in [a, b]$. There are several approaches to obtaining calculation formulas for this method [1].

We replace the region of continuous variation of the argument x with a discrete set of points $x_j = a + jh$, called grid nodes:

$$h = \frac{b-a}{N}, \quad x_0 = a, \quad x_N = b,$$

where h is the grid step.

Let us write eq. (1.24) for $x = x_j$. By definition of the derivative, we have

$$\left.\frac{dy}{dx}\right|_{x=x_j} = \lim_{h\to 0} \frac{y(x_j + h) - y(x_j)}{h}. \tag{1.26}$$

Discarding the limit in eq. (1.26), we replace the derivative with a finite-difference relation:

$$\left.\frac{dy}{dx}\right|_{x=x_j} \approx \frac{y(x_j + h) - y(x_j)}{h}. \tag{1.27}$$

Substituting (1.27) into (1.24), we obtain:

$$\frac{y(x_j + h) - y(x_j)}{h} \approx f(x_j, y(x_j)), \; j = 0, 1, \ldots, N-1. \tag{1.28}$$

where $y(x_i)$ is the value of the solution $y(x)$ of the Cauchy problem at node x_i. Denote by y_j the numerical solution satisfying the difference equation:

$$\frac{y_{j+1}-y_j}{h} \approx f(x_j,y_j),\ j=0,1,\ldots,N-1 \tag{1.29}$$

Or

$$y_{j+1}=y_j+hf(x_j,y_j),\quad j=0,1,\ldots,N-1. \tag{1.30}$$

Considering the initial condition (1.25), using the difference equation (1.30), we can sequentially determine

$$\begin{aligned} y_1 &= y_0 + hf(x_0,y_0), && (j=0),\\ y_2 &= y_1 + hf(x_1,y_1), && (j=1),\\ y_N &= y_{N-1} + hf(x_{N-1},y_{N-1}), && (j=N-1). \end{aligned}$$

1.3.2 Implicit Euler method

In the implicit Euler method, the calculation is carried out according to the formula:

$$y_{j+1}=y_j+\frac{h}{2}\left[f(x_j,y_j)+f(x_{j+1},y_{j+1})\right],\quad j=0,1,\ldots,N-1. \tag{1.31}$$

Unlike (1.30), the difference scheme (1.31) does not allow explicitly expressing y_{j+1} through y_j. Scheme (1.31) is implicit. To solve the difference equation, it is necessary to apply one or another iterative process, using y_i as the initial approximation.

1.3.3 Improved explicit Euler method

In the improved explicit Euler method, the calculation is carried out according to the formula:

$$y_{j+1}=y_j+\frac{h}{2}\left[f(x_j,y_j)+f(x_{j+1},y_j+hf(x_j,y_j))\right],\quad j=0,1,\ldots,N-1. \tag{1.32}$$

According to the formula, the calculation is performed explicitly.

1.4 Numerical methods for a parabolic equation

1.4.1 Explicit difference scheme

Let us consider a one-dimensional differential equation of parabolic type with initial and boundary conditions:

$$\frac{\partial u}{\partial t}=\sigma\frac{\partial^2 u}{\partial x^2}+f(u,t,x); \qquad u\,(t=0,x)=\xi\,(x); \qquad \begin{cases} u(t,x=a)=\varphi_1(t); \\ u(t,x=b)=\varphi_2(t). \end{cases} \tag{1.33}$$

Let us write an explicit difference scheme for eq. (1.33):

$$\frac{u_j^{n+1}-u_j^n}{\Delta t}=\sigma\frac{u_{j+1}^n-2u_j^n+u_{j-1}^n}{h^2}+f(u_j^n,t^n,x_j). \tag{1.34}$$

Let us write an approximation of the initial and boundary conditions:

$$u_j^0=\xi\,(x_j); \qquad u_1^{n+1}=\varphi_1(t^{n+1}); \qquad u_N^{n+1}=\varphi_2(t^{n+1}). \tag{1.35}$$

The approximation of the boundary conditions (1.35) is written at the $(n+1)$th time step for the convenience of the subsequent presentation of the method and algorithm for solving the explicit difference scheme (1.34).

To solve the problem using the explicit scheme, we express the quantity u_j^{n+1} from the difference scheme (1.34):

$$u_j^{n+1}=u_j^n+\frac{\sigma\,\Delta t}{h^2}\,(u_{j+1}^n-2u_j^n+u_{j-1}^n)+f(u_j^n,\ t^n,\ x_j)\,\Delta t. \tag{1.36}$$

The values u_j^0 for any j can be determined using the approximation of the initial condition (1.35). If we set $n=0$, then using expression (1.36), we can determine the values u_j^1, $j=2,\ \ldots,\ N-1$. The values u_1^1, u_N^1 are determined using the approximation of the boundary conditions (1.35). Next, we set $n=1$ and from expression (1.36), we determine u_j^2, $j=2,\ \ldots,\ N-1$, and the values u_1^2, u_N^2 again using the approximation of the boundary conditions (1.35), and so on. Thus, relation (1.36) allows us to calculate all the values of the desired function u at the nodes of the difference grid, except for the values defined using the initial and boundary conditions. The described technique for solving an explicit difference scheme is clearly characterized by its difference template (see the figure), in which the nth time step should be considered known, and the $(n+1)$th time step – the desired one.

Relations of type (1.36), which allow one to calculate the values of the sought function u at the nodes of the difference grid through the known values of the function u at other (usually neighboring) nodes of the difference grid, are called recurrence relations.

1.4.2 Implicit difference scheme

Let us consider a one-dimensional differential equation of parabolic type with initial and boundary conditions:

$$\frac{\partial u}{\partial t} = \sigma \frac{\partial^2 u}{\partial x^2} - k\,u + f(t,x), \qquad k \ge 0; \quad u\,(t=0,x) = \xi\,(x); \quad \begin{cases} u(t,x=a) = \varphi_1(t); \\ u(t,x=b) = \varphi_2(t). \end{cases} \tag{1.37}$$

Let us write an implicit difference scheme for eq. (4.5):

$$\frac{u_j^{n+1} - u_j^n}{\Delta t} = \sigma \frac{u_{j+1}^{n+1} - 2u_j^{n+1} + u_{j-1}^{n+1}}{h^2} - k\,u_j^{n+1} + f(t^n,\ x_j). \tag{1.38}$$

Let us write an approximation of the initial and boundary conditions:

$$u_j^0 = \xi\,(x_j); \qquad u_1^{n+1} = \varphi_1(t^{n+1}); \qquad u_N^{n+1} = \varphi_2(t^{n+1}). \tag{1.39}$$

The approximation of the boundary conditions (1.39) is written at the $(n+1)$th time step for the convenience of the subsequent presentation of the method and algorithm for solving the implicit difference scheme (1.38).

The explicit difference scheme has a fairly simple solution method, and the method for solving the implicit difference scheme (1.38), called the sweep method, is more complex. Before moving on to the presentation of the sweep method, it is necessary to derive a number of relationships used by this method.

We transform the implicit difference scheme (1.38), grouping the terms containing the value of the function u at the $(n+1)$th time step on the left-hand side, and all the remaining terms on the right-hand side:

$$-\frac{\sigma\,\Delta t}{h^2} u_{j+1}^{n+1} + \left(1 + 2\frac{\sigma\,\Delta t}{h^2} + k\,\Delta t\right) u_j^{n+1} - \frac{\sigma\,\Delta t}{h^2} u_{j-1}^{n+1} = u_j^n + \Delta t\,f(t^n, x_j). \tag{1.40}$$

We introduce the following notations:

$$a_j = -\frac{\sigma\,\Delta t}{h^2}; \quad b_j = 1 + 2\frac{\sigma\,\Delta t}{h^2} + k\,\Delta t; \quad c_j = -\frac{\sigma\,\Delta t}{h^2}; \quad \xi_j^n = u_j^n + \Delta t\,f(t^n,\ x_j). \tag{1.41}$$

Taking into account the notations (1.41), equality (1.40) will have the form:

$$a_j\,u_{j+1}^{n+1} + b_j\,u_j^{n+1} + c_j\,u_{j-1}^{n+1} = \xi_j^n. \tag{1.42}$$

This transformation is often called the transformation of the implicit difference scheme to a form convenient for using the Tomas algorithm.

1.4.3 Crank-Nicholson difference scheme

Consider a one-dimensional differential equation of parabolic type with initial and boundary conditions:

$$\frac{\partial u}{\partial t} = \sigma \frac{\partial^2 u}{\partial x^2} - k\,u + f(t,x), \qquad k \geq 0; \quad u\,(t=0,x) = \xi\,(x); \qquad \begin{cases} u(t,x=a) = \varphi_1(t); \\ u(t,x=b) = \varphi_2(t). \end{cases} \tag{1.43}$$

Write a difference scheme for eq. (1.43), representing the second derivative of the function u with respect to the coordinate x as two terms:

$$\frac{\partial^2 u}{\partial x^2} = \frac{1}{2}\frac{\partial^2 u}{\partial x^2} + \frac{1}{2}\frac{\partial^2 u}{\partial x^2}$$

and approximating the first term at the nth time step, and the second at the $(n+1)$th time step:

$$\frac{u_j^{n+1} - u_j^n}{\Delta t} = \frac{\sigma}{2}\frac{u_{j+1}^n - 2u_j^n + u_{j-1}^n}{h^2} + \frac{\sigma}{2}\frac{u_{j+1}^{n+1} - 2u_j^{n+1} + u_{j-1}^{n+1}}{h^2} - k\,u_j^{n+1} + f(t^n, x_j). \tag{1.44}$$

This difference scheme is called the Crank-Nicholson difference scheme.

1.5 Solution of the Dirichlet problem for the Laplace equation using the grid method

Find a continuous function $u(x, y)$ satisfying the Laplace equation inside a rectangular domain:

$$\Delta u \equiv \frac{\partial^2 u}{\partial x^2} + \frac{\partial^2 u}{\partial y^2} = 0$$

and taking on the boundary of the domain Ω the given values, that is:

$$\begin{aligned} u(0,y) &= f_1(y),\ y \in [0,b], \qquad & u(a,y) &= f_2(y),\ y \in [0,b], \\ u(x,0) &= f_3(x),\ x \in [0,a], \qquad & u(x,b) &= f_4(x),\ x \in [0,a], \end{aligned}$$

where f_1, f_2, f_3, f_4 are the given functions.

We will assume that $u(x, y)$ is continuous on the boundary of the domain Ω, that is, $f_1(0) = f_3(0)$, $f_1(b) = f_4(0)$, $f_2(0) = f_3(a)$, and $f_2(b) = f_4(a)$. Having chosen the steps h, in x and y, respectively, we construct the grid $x_i = ih$, $i = 0,1,\ldots,n$, $y_j = jl$, and $j = 0,1,\ldots,m$, where $x_n = nh = a$ and $y_m = ml = b$.

Introducing the notations $u_{ij} = u(x_i, y_j)$, we approximate the partial derivatives $\frac{\partial^2 u}{\partial x^2}$ and $\frac{\partial^2 u}{\partial y^2}$ at each internal node of the grid by the central-difference derivatives of the second order:

$$\frac{\partial^2 u}{\partial x^2} = \frac{u_{i+1,j} - 2u_{ij} + u_{i-1,j}}{h^2} + O(h^2),$$

$$\frac{\partial^2 u}{\partial y^2} = \frac{u_{i,j+1} - 2u_{ij} + u_{i,j-1}}{l^2} + O(l^2),$$

and replace the Laplace equation with the finite-difference equation:

$$\frac{u_{i+1,j} - 2u_{ij} + u_{i-1,j}}{h^2} + \frac{u_{i,j+1} - 2u_{ij} + u_{i,j-1}}{l^2} = 0, i = 1, \ldots, n-1, j = 1, \ldots, m-1. \tag{1.45}$$

The error in replacing the differential equation with a difference equation is $O(h^2) + O(l^2)$.

Equation (1.45) together with the values u_{ij} at the boundary nodes forms a system of linear algebraic equations for approximate values of the function $u(x,y)$ at the grid nodes (x_i, y_j). This system has its simplest form at $l = h$:

$$u_{ij} = (u_{i+1,j} + u_{i-1,j} + u_{i,j+1} + u_{i,j-1})/4 \tag{1.46}$$

$$u_{i0} = f_3(x_i),\ u_{im} = f_4(x_i),\ u_{0j} = f_1(y_j),\ u_{nj} = f_2(y_j),$$

$$i = 1, \ldots, n-1, \quad j = 1, \ldots, m-1.$$

When obtaining the grid equation (1.46), the node scheme shown in Fig. 1.1 was used. The set of nodes used to approximate the equation at a point is called a template. In this chapter, a "cross"-type template is used.

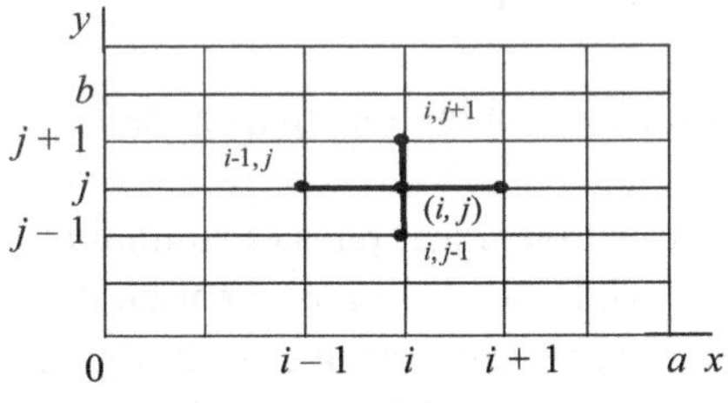

Fig. 1.1: The node scheme.

The numerical solution of the Dirichlet problem for the Laplace equation in a rectangle consists of finding approximate values of the desired function $u(x, y)$ at the internal nodes of the grid. To determine the quantities, it is necessary to solve the system of linear algebraic equation (1.46). There are various methods for solving eq. (1.46).

Chapter 2
The method of moving nodes for two-point boundary value problems

This chapter introduces the concept of a moving node, and also provides approximate solutions of simple problems using a moving node for two-point boundary value problems.

2.1 The concept of a moving node

Based on the previous chapter, the numerical solution of two-point boundary value problems for ordinary differential equations (ODEs) is reduced to an approximation of the derivatives of the equation and the boundary conditions. The resulting discrete algebraic equation is most often solved by the sweep method. Thus, we obtain approximate values of the solution at the nodal points, that is, we have a solution in the form of a table.

When applying the finite-difference method (FDM) to solving a differential equation (DE), there is a transition from a continuous domain to a finite-difference one. A grid of "nodal points" is introduced in the solution domain. Representing the derivatives in finite-difference form, it is reduced to the form of a difference equation. The difference equation is written for all grid nodes and results in a system of algebraic equations [1, 22, 25, 27, 28].

Discrete method, in particular, the FDM), also allows obtaining approximate analytical solutions to boundary value problems.

Most of the DEs encountered in equations of mathematical physics contain only partial derivatives of the first and second orders, while for the approximation of derivatives they try to use no more than three nodes of the difference grid (in the case of ODEs) (Fig. 2.1). Let the node W and E be considered fixed (i.e., boundary conditions), and the node x changes to segments (W, E). Then the approximation of the derivatives (first or second order) also changes based on the location of the internal node. The node x is called movable [31].

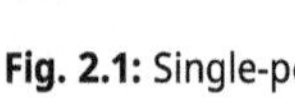

Fig. 2.1: Single-point movable node.

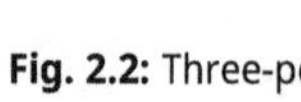

Fig. 2.2: Three-point movable node.

https://doi.org/10.1515/9783111308173-003

Knowing the boundary values at an arbitrary point x, we obtain an approximate value of the desired function at any point of the segment $[W, E]$. Thus, the analytical expression for the unknown function is determined [2, 3].

It is possible to increase the number of movable nodes. We will select additional movable nodes as follows: $x_1 = (W + x)/2$, $x_2 = (E + x)/2$. When node x changes its position, x_1 and x_2 automatically change their positions (Fig. 2.2). In this way, it is possible to increase the number of movable nodes. Increasing the number of movable nodes is related to the accuracy of difference equations.

Moving nodal points applies not only to finite-difference equations; this approach can be successfully applied to discretization of DEs using the control volume method [25].

2.2 Obtaining an approximate analytical solution of an ODE with one movable node

2.2.1 The concept of a movable node

We need to find a solution of an ODE in a domain with appropriate boundary conditions. Let us take an arbitrary point. We have three nodes: two boundary nodes and an interior node (Fig.2.1). The position of the point inside the domain is determined by the movable node x. The difference equation is usually written for an arbitrary node. When approximating differential operators, the first derivatives at the movable node are approximated by difference relations:

$$\frac{dU}{dx} \approx \frac{U(x) - U(W)}{x - W}, \tag{2.1}$$

$$\frac{dU(x)}{dx} \approx \frac{U(E) - U(x)}{E - x}, \tag{2.2}$$

$$\frac{dU(x)}{dx} \approx \frac{U(E) - U(W)}{E - W}. \tag{2.3}$$

Approximation of the derivative by eq. (2.1) is called approximation of this derivative using backward differences, approximation of eq. (2.2) using forward differences, and eq. (2.3) is approximation using a central difference.

The second derivative on a moving node is approximated as follows [25, 27] (similar to the approximation of the second derivative in a nonuniform grid).

We write the second derivative as follows:

$$\frac{d^2U(x)}{dx^2} = \frac{d}{dx}\left(\frac{dU}{dx}\right)$$

We approximate the first derivative on the segments $[x_1, x_2]$:

$$\frac{d^2U(x)}{dx^2} = \frac{d}{dx}\left(\frac{dU}{dx}\right) \approx \left(\frac{\left(\frac{dU}{dx}\right)_{x_2} - \left(\frac{dU}{dx}\right)_{x_1}}{x_2 - x_1}\right)$$

We replace the first derivative with the central difference, we have

Since $x_2 - x_1 = \frac{E-W}{2}$, for the approximation of the second derivative we obtain the expression:

$$\frac{d^2U(x)}{dx^2} \approx \frac{2}{E-W}\left(\frac{U(E)-U(x)}{E-x} - \frac{U(x)-U(W)}{x-W}\right) \tag{2.4}$$

We consider some model problems of applying the moving nodes method (MNM) to obtain an analytical solution.

2.2.2 Boundary value problem with diffusion

Let us consider the simplest ODE with boundary conditions:

$$\frac{d^2u}{dx^2} = C, \quad u(0) = 0, \quad u(1) = 0, \tag{2.5}$$

where C is a constant. In our example $W = 0$, $E = 1$. Let x be an arbitrary point on the segments $[W,E]$. We approximate the second derivative in eq. (2.5) using eq. (2.4). As a result, we obtain one algebraic equation:

$$\frac{2}{E-W}\left(\frac{U(1)-U(x)}{1-x} - \frac{U(x)-U(0)}{x-0}\right) = C. \tag{2.6}$$

In eq. (2.6), $U(x)$ is an approximate solution to problem (2.5).

Solving eq. (2.6) for $U(x)$, we obtain:

$$U(x) = -\frac{C}{2}x(1-x). \tag{2.7}$$

It is easy to obtain an exact solution to problem (2.5):

$$u(x) = -\frac{C}{2}x(1-x). \tag{2.8}$$

For this problem, the exact solution coincides with the solution of the obtained by MNM. This means that for the problem under consideration, the approximation of the second derivative by formula (2.4) is exact.

However, it is not always possible to obtain exact approximations of the derivatives.

2.2.3 MNM for the convection-diffusion equation

Consider the transport equation

$$\frac{d\Phi}{dx} = \frac{1}{Pe}\frac{d^2\Phi}{dx^2} + S(x). \tag{2.9}$$

Here the unknown function, the source, is the Peclet number. The equation is considered under the corresponding boundary conditions.

We approximate the convective term of eq. (2.9) by eq. (2.1) and the diffusion term by eq. (2.4). We consider eq. (2.9) on segments [0, 1] with boundary conditions: $\Phi(0) = 0$, $\Phi(1) = 1$. Then, using the upwind scheme, we replace eq. (2.9) with a difference equation, which looks like this:

$$\frac{U(x) - U(0)}{x - 0} = \frac{2}{Pe}\left(\frac{U(1) - U(x)}{1 - x} - \frac{U(x) - U(0)}{x - 0}\right). \tag{2.10}$$

In eq. (2.10), $U(x)$ is an approximate solution at point x. Considering that $\Phi(0) = U(0)$, $\Phi(1) = U(1)$, from eq. (2.10), taking into account the boundary conditions, we easily determine the approximate solution $U(x)$:

$$U(x) = \frac{2x}{2 + Pe(1 - x)}. \tag{2.11}$$

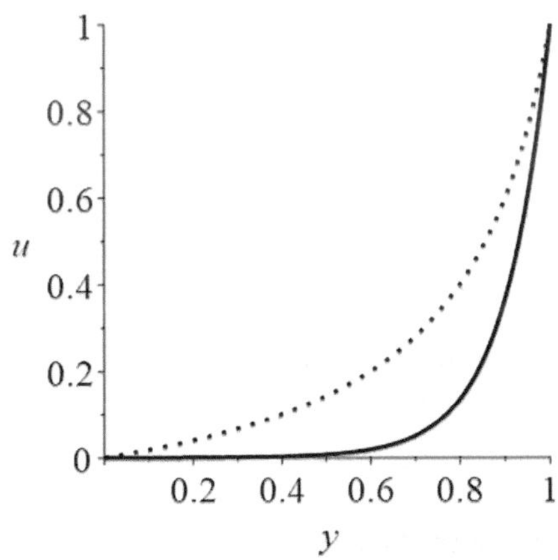

Fig. 2.3: Comparison of the exact and approximate solutions ($Pe = 10$).

Figure 2.3 shows a comparison of the exact and approximate solutions. The solid line corresponds to the exact solution and the dotted line to the solution (2.11). It is clear from the graph that there is numerical diffusion.

For $\Phi(0) = 0$, $\Phi(1) = 1$ and $S(x) = 5\cos 4x$, $Pe = 5$ the results of the exact and approximate solutions are shown in Fig. 2.4. It is clear from the graph that large errors occur.

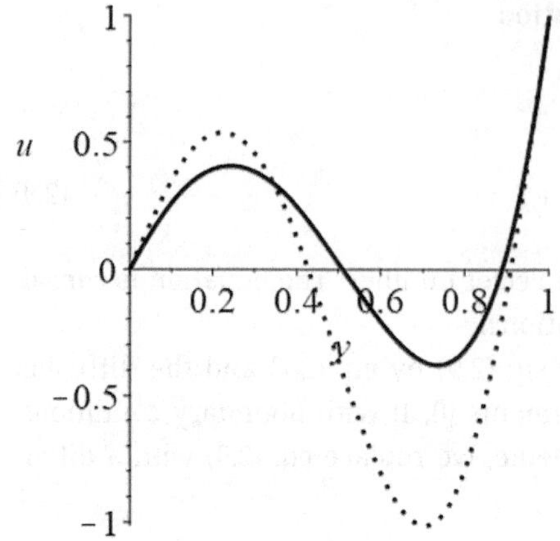

Fig. 2.4: Comparison of exact and approximate solutions (*Pe* = 5).

The Peclet number plays a significant role here. Indeed, for Pe=0.1, we obtain the solutions shown in Fig. 2.5, which shows that the approximate and exact solutions are close.

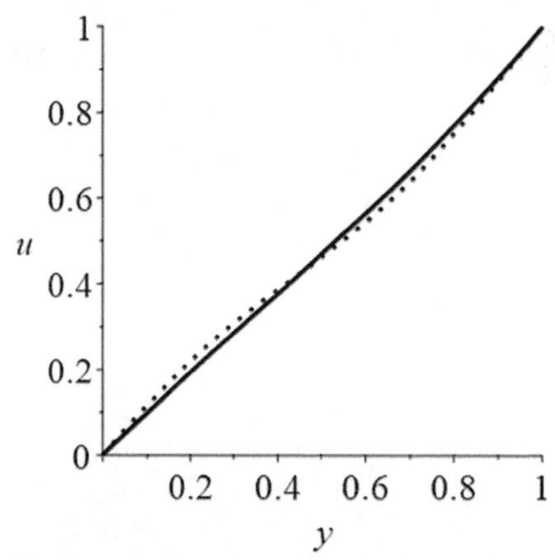

Fig. 2.5: Comparison of exact and approximate solutions (*Pe* = 0.1).

2.2.4 Convection-diffusion equation with a variable coefficient

Let us consider the equation [9]

$$\varepsilon \frac{d^2u}{dx^2} + 2x \frac{du}{dx} = 0 \tag{2.12}$$

into segments (−1,1) with the boundary conditions $u(1) = -1$, $u(1) = 2$. The exact solution is determined through the error functions:

$$u = \frac{\operatorname{erf}(1/\sqrt{\varepsilon}) + 3\operatorname{erf}(x/\sqrt{\varepsilon})}{2\operatorname{erf}(1/\sqrt{\varepsilon})} - 1.$$

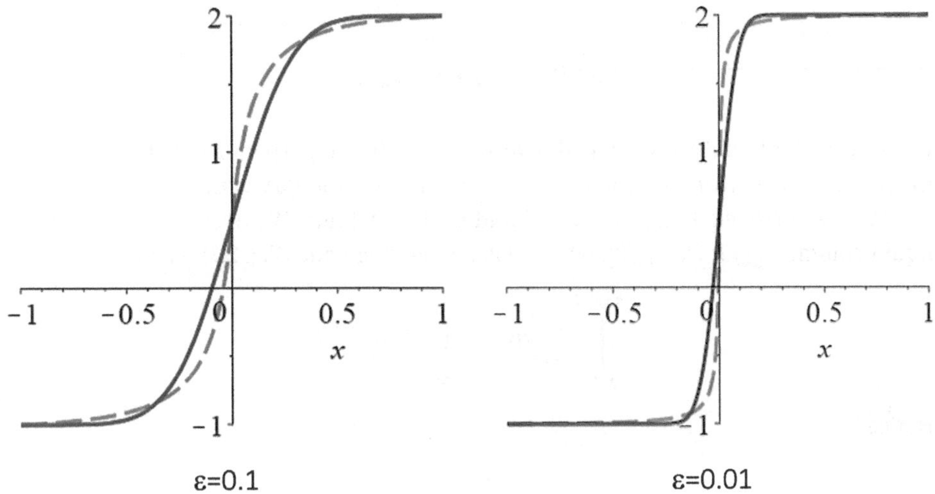

Fig. 2.6: Comparison of solutions.

The difference scheme with a moving node for eq. (2.12) has the form (upwind scheme):

$$\varepsilon\left[\frac{2-U(x)}{1-x} - \frac{U(x)+1}{x+1}\right] + \frac{1}{2}(2x-|2x|)\frac{U(x)+1}{1+x} + \frac{1}{2}(2x+|2x|)\frac{2-U(x)}{1+x} = 0.$$

By solving this equation with respect to $U(x)$, we obtain an approximate analytical solution. Figure 2.6 shows a comparison of the solutions of the exact and approximate analytical solutions; the solid line corresponds to the exact solution and the dotted line to the approximate one.

Let us now consider several examples of an applied nature.

2.2.5 Flow in a flat pipe

The flow of a viscous liquid in a flat pipe in a one-dimensional formulation is described by the equation [22]:

$$\frac{d^2U}{dy^2} = -\frac{\Delta p}{\mu\, l}, \tag{2.13}$$

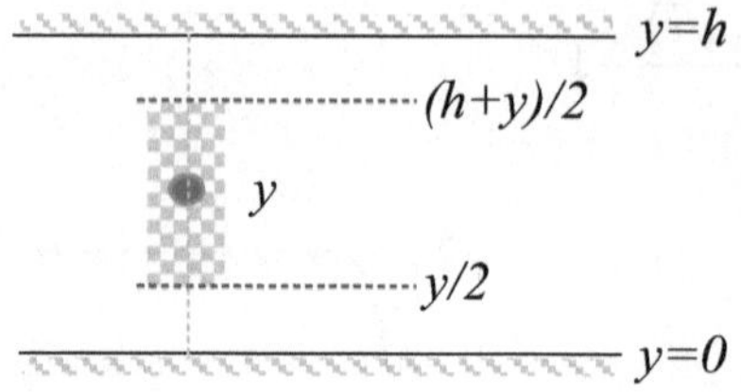

Fig. 2.7: Averaging area.

where U is the liquid velocity, is the vertical coordinate perpendicular to the flow, is the pressure drop (const) is the viscosity. Let and be stationary walls.

We discretize eq. (2.13) over the liquid control volume. We average (2.13) over the liquid volume $-[y/2, (h-y)/2]$, where y is the moving node (Fig. 2.7). Then we have

$$\int_{y/2}^{(h+y)/2} \frac{d^2U}{dy^2} dy = \int_{y/2}^{(h+y)/2} \left(-\frac{\Delta p}{\mu\, l}\right) dy$$

Hence

$$\left.\frac{dU}{dy}\right|_{(h+y)/2} - \left.\frac{dU}{dy}\right|_{y/2} = \left(-\frac{\Delta p}{\mu\, l}\right)\frac{h}{2}.$$

We replace the derivatives with a difference ratio:

$$\left.\frac{dU}{dy}\right|_{(h+y)/2} \approx \frac{u(h)-u(y)}{h-y}, \quad \left.\frac{dU}{dy}\right|_{y/2} \approx \frac{u(y)-u(0)}{y}.$$

Here the approximate value is $u(y)$. Thus, approximation (2.13) by the moving node has the form:

$$\frac{u(h)-u(y)}{h-y} - \frac{u(y)-u(0)}{y} = \left(-\frac{\Delta p}{\mu\, l}\right)\frac{h}{2}. \tag{2.14}$$

Hence, taking into account the no-slip condition ($u(h) = u(0) = 0$):

$$u(y) = -\frac{\Delta p}{2\mu\, l} y(h-y).$$

Here $u(y)$ is the averaged solution. For this problem, the averaged solution coincides with the exact solution [21].

This indicates that approximation (2.16) for this problem is exact. The reason for the coincidence of the solution obtained using the MNM with one node, and the exact solution is explained by the following fact.

Lagrange's mean value theorem states that if a function $f(x)$ is continuous on a segment $[a, b]$ and differentiable on an interval, then in this interval, there exists at least one point $x = \xi$, such that

$$\frac{f(b)-f(a)}{b-a} = f'(\xi). \tag{2.15}$$

It is easy to verify that if $f(x)$ represents a parabola, then in eq. (2.15), $\zeta = (a+b)/2$. The exact solution (2.13) is a parabola. Integrating eq. (2.13) over the control volume, we obtain

$$\int_{y/2}^{(h+y)/2} \frac{d^2u}{dy^2}dy = \left.\frac{du}{dy}\right|_{(h+y)/2} - \left.\frac{du}{dy}\right|_{y/2} = \int_{y/2}^{(h+y)/2} \frac{1}{\mu}\frac{\Delta p}{l}dy.$$

Since $u(y)$ there is a parabola, therefore

$$\left.\frac{du}{dy}\right|_{(h+y)/2} = \frac{u(h)-u(y)}{h-y}, \quad \left.\frac{du}{dy}\right|_{y/2} = \frac{u(y)-u(0)}{y-0},$$

and eq. (2.14) is an exact difference analog of eq. (2.13).

2.2.6 Flow in a round pipe

The steady-state flow of a viscous fluid in a round pipe in a one-dimensional formulation is described by the equation [22]:

$$\frac{1}{r}\frac{d}{dr}\left(r\frac{dU}{dr}\right) = -\frac{\Delta p}{\mu l}, \tag{2.16}$$

where U is the fluid velocity, r is the radial coordinate perpendicular to the flow, $\Delta p/l$ is the pressure difference (const), and μ is the viscosity. In the cylindrical coordinate system, $r=0$ corresponds to the axis of the cylinder and the radius of the pipe.

The exact solution of eq. (2.16) with the no-slip condition has the form:

$$u(r) = \frac{A}{4}(R^2 - r^2),$$

where $A = \Delta P/(\mu \cdot l)$.

We integrate eq. (2.16) into the segments $[r_1, r_2]$, $r_1 = r/2$, $r_2 = (R+r)/2$:

$$\int_{r_1}^{r_2} \frac{d}{dr}\left(r\frac{dU}{dr}\right)dr = -A\int_{r_1}^{r_2} r\,dr.$$

Hence

$$r\left.\frac{dU}{dr}\right|_{r=r_2} - r\left.\frac{dU}{dr}\right|_{r=r_{r1}} = -A\left(\frac{r_2^2}{2} - \frac{r_1^2}{2}\right)$$

or

$$\frac{R+r}{2}\frac{dU}{dr}\bigg|_{r=r_2} - \frac{r}{2}\frac{dU}{dr}\bigg|_{r=r_{r1}} = -\frac{A}{2}\left(\left(\frac{R+r}{2}\right)^2 - \left(\frac{r}{2}\right)^2\right).$$

Replace the derivatives with central differences:

$$\frac{R+r}{2}\,\frac{U(R)-U(r)}{R-r} - \frac{r}{2}\,\frac{U(r)-U(0)}{r-0} = -\frac{A}{2}\left(\left(\frac{R+r}{2}\right)^2 - \left(\frac{r}{2}\right)^2\right). \tag{2.17}$$

From here we determine $U(r)$

$$U(r) = \frac{1}{2}\frac{R-r}{R}U(0) - \frac{A}{2}\frac{R-r}{R}\left(\left(\frac{R+r}{2}\right)^2 - \left(\frac{r}{2}\right)^2\right) \tag{2.18}$$

We determine the velocity at the point 0 as follows:

In eq. (2.17) we pass to the limit at $r \to 0$, then we have

$$-\frac{U(0)}{2} = -\frac{A}{2}\left(\frac{R}{2}\right)^2.$$

From here we obtain the values of the velocity at the point 0:

$$U(0) = A\left(\frac{R}{2}\right)^2.$$

We put the obtained expression in eq. (2.18) and obtain a solution in the form:

$$U(r) = \frac{A}{4}(R^2 - r^2),$$

which coincides with the exact solution.

2.2.7 Heat propagation in a plate

Heat propagation in a plate is described by equation [1]:

$$\frac{d^2T}{dx^2} + \frac{q}{k} = 0, \qquad \frac{dT(0)}{dx} = 0, \qquad T(1) = 1, \tag{2.19}$$

where k is the specific thermal conductivity, and q is the heat release per unit volume (k and q = const). It is assumed that the source does not depend on temperature. Replacing eq. (2.18) with a difference equation with a moving node we have

$$\frac{2}{1-0}\left[\frac{T(1)-T(x)}{1-x}-\frac{T(x)-T(0)}{x-0}\right]+\frac{q}{k}=0. \tag{2.19}$$

Solving equation from here, we obtain

$$T(x)=1+\frac{q}{2k}(1-x^2). \tag{2.20}$$

Solution (2.20) coincides with the exact solution. Note that the exact solution is obtained not only for the Dirichlet problem as in the problem of flow in a flat pipe. Here the boundary conditions are of mixed type.

2.2.8 Magnetohydrodynamic Couette flow

Let us consider the Couette flow, when a conducting fluid flows in a uniform magnetic field between two plates, one of which is stationary, and the other moves in its own plane with a constant velocity. Based on the Navier-Stokes equation, taking into account the magnetic field and taking into account the one-dimensionality of the flow in dimensionless form, we can write as follows [23]:

$$\frac{d^2u}{dy^2}-M^2u=P. \tag{2.21}$$

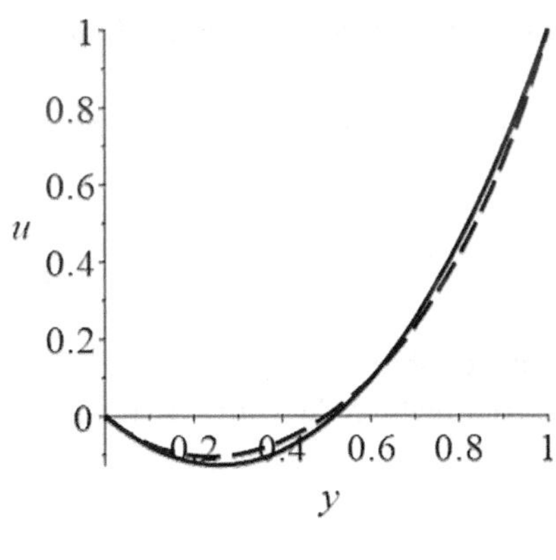

Fig. 2.8: Comparison of the exact and approximate solutions ($M = 2$, $P = 4$). The solid line is the exact solution, and the dotted line is according to eq. (2.20).

Boundary conditions

$$u(0)=0, \quad u(1)=1. \tag{2.22}$$

Here u is the dimensionless flow velocity and y is the dimensionless coordinate. Dimensionless quantities M are the Hartmann number and P the pressure coefficient (M and P = const).

Replacing the second-order derivative in eq. (2.21) with a difference relation similar to eq. (2.4), and taking into account the boundary condition (2.22), we can obtain an approximate solution:

$$u_1(y) = \frac{2y - Py(1-y)}{2 + M^2 y(1-y)}. \tag{2.23}$$

This solution comes close to the exact solution (Fig. 2.8).

Remark 1. In the examples given, the convective term is approximated by an upwind scheme. Other approximations can be used for improvement.

Remark 2. In the examples given, the approximation of the term with the source is carried out by a constant in the considered moving segment. For improvement, other approximations can be used to obtain an improved solution.

2.3 Obtaining an analytical solution with several moving nodes

The previous examples showed that using one moving node in most cases, more rough solutions are obtained. In order to improve the analytical solution, the idea of increasing the number of moving nodes should be used.

2.3.1 Movable node method for a one-dimensional convection-diffusion problem

Due to the importance of convection-diffusion problems, we will apply the multipoint MNM to the convection-diffusion equation:

$$\frac{d\Phi}{dx} = \frac{1}{Pe}\frac{d^2\Phi}{dx^2} + S(x). \tag{2.24}$$

Inside the interval, we take an arbitrary single node $x \in (W, E)$.

Let us consider the difference analog of eq. (2.24), in which the convective term is approximated by a one-sided difference ratio [3].

Then the upstream difference scheme has the form:

$$Pe\frac{U^1 - U_W^1}{x - W} = \frac{2}{(E-W)}\left(\frac{U_E^1 - U^1}{E - x} - \frac{U^1 - U_W^1}{x - W}\right) + Pe \cdot S(x). \tag{2.25}$$

In eq. (2.25), $U^1 = U^1(x)$ means an approximate solution of Φ when choosing one movable node. Scheme (2.25) can be rewritten as follows:

$$a_P^1 U^1 = a_E^1 U_E^1 + a_W^1 U_W^1 + F^1(x), \tag{2.26}$$

Here

$$a_E^1 = \frac{2}{(E-W)(E-x)}, \; a_W^1 = \frac{Pe}{(x-W)} + \frac{2}{(E-W)(x-W)}, \; a_P^1 = a_E^1 + a_W^1,$$

$$F^1(x) = Pe \cdot S(x).$$

Hence we have

$$U^1 = \frac{2(x-W)U_E^1 + (E-x)(2+Pe(E-W))U_W^1}{(E-W)(2+Pe(E-x))} + \frac{(x-W)(E-x)}{2+Pe(E-x)} Pe \cdot S(x). \tag{2.27}$$

When $x \in (W,E)$ changes its position (let us make it movable inside the interval (W,E)), based on eq. (2.27), we obtain the values of the unknown function at each position. In other words, obtained with the help of eq. (2.27), it will give us an approximate solution of the problem. Note that in this case the upper index corresponds to the number of movable nodes $U_W^1 = \Phi(W), \; U_E^1 = \Phi(E)$.

Add additional movable nodes

$$x_1 = \frac{x+W}{2}, x_2 = \frac{x+E}{2}.$$

Now we have three movable nodes $x, \; x_1, \; x_2$. Note that if x changes its positions, then x_1 and x_2. also change their positions.

A scheme of type (2.27) for the segment $[W,x]$ has the form:

$$Pe \frac{U_1^3 - U_W^3}{(x-W)/2} = \frac{2}{(x-W)} \left(\frac{U^3 - U_1^3}{x - x_1} - \frac{U_1^3 - U_W^3}{x_1 - W} \right) + Pe \cdot S(x_1). \tag{2.28}$$

Here $U_1^3 = U^3(x_1)$. A scheme of type (2.27) for the segment $[x,E]$ has the form:

$$Pe \frac{U_2^3 - U^3}{(E-x)/2} = \frac{2}{(E-x)} \left(\frac{U_E^3 - U_2^3}{E - x_2} - \frac{U_2^3 - U^3}{x_2 - x} \right) + Pe \cdot S(x_2). \tag{2.29}$$

A scheme against the flow for the segment $[x_1, x_2]$:

$$Pe \frac{U^3 - U_1^3}{x - x_1} = \frac{2}{(x_2 - x_1)} \left(\frac{U_2^3 - U^3}{x_2 - x} - \frac{U^3 - U_1^3}{x - x_1} \right) + Pe \cdot S(x). \tag{2.30}$$

Here $U_2^3 = U^3(x_2)$.

In eq. (2.30) we exclude $U_1^3, \; U_2^3$ using eqs. (2.28) and (2.29). Then we get the following scheme:

$$Pe \frac{U^3 - U_W^3}{\frac{(x-W)}{2} \cdot (1+\tau_1)} = \frac{4}{(E-W)} \left(\frac{U_E^3 - U^3}{\frac{E-x}{2} \cdot (1+\gamma_1)} - \frac{U^3 - U_W^3}{\frac{x-W}{2} \cdot (1+\tau_1)} \right) + F^3(x). \tag{2.31}$$

Here we use the notation:

$$\tau_1 = 2/(2+\sigma),\ \gamma_1 = (2+\theta)/2,\ \sigma = Pe\,(x-W),\ \theta = Pe\,(E-x),$$

$$F^3(x) = Pe\cdot S(x) + \frac{4+Pe\cdot(E-W)}{E-W}\cdot\frac{1-\tau_1}{1+\tau_1}\cdot S(x_1) + \frac{4}{E-W}\cdot\frac{\gamma_1-1}{\gamma_1+1}\cdot S(x_2),$$

$$U_W^3 = \Phi(W),\ U_E^3 = \Phi(E)$$

is introduced.

Equation (2.31) can be rewritten as follows:

$$a_P^3 U^3 = a_E^3 U_E^3 + a_W^3 U_W^3 + F^3(x), \tag{2.32}$$

where

$$a_E^3 = \frac{8}{(E-W)(E-x)(1+\gamma_1)},\ a_W^3 = \frac{2Pe}{(x-W)(1+\tau_1)} + \frac{8}{(E-W)(x-W)(1+\tau_1)},$$

$$a_P^3 = a_W^3 + a_E^3.$$

We increase the number of movable nodes:

$$x_1^- = \frac{x_1+W}{2} = \frac{x+3W}{4},\ x_1^+ = \frac{x_1+x}{2} = \frac{3x+W}{4},\ x_2^- = \frac{x_2+x}{2} = \frac{3x+E}{4},$$

$$x_2^+ = \frac{x_2+E}{2} = \frac{x+3E}{4}.$$

In the difference scheme (2.32), the unknown function appears in three nodes: W, x, and E. The function S is calculated at the points x_1, x, x_2. Let us write a scheme of type (2.32) for each of the segments $[W, x]$, $[x, W]$, and $[x_1, x_2]$.

A scheme of type (2.32) for a segment $[W,x]$ has the form:

$$a_{x_1}^3 U_{x_1}^3 = a_x^3 U_x^3 + a_{W-}^3 U_W^3 + F_-^3(x_1), \tag{2.33}$$

where

$$a_x^3 = \frac{8}{(x-W)(x-x_1)(1+\gamma_1^-)},\ a_{W-}^3 = \frac{2Pe}{(x_1-W)(1+\tau_1^-)} + \frac{8}{(x-W)(x_1-W)(1+\tau_1^-)},$$

$$a_{x_1}^3 = a_x^3 + a_{W-}^3,$$

$$F_-^3(x_1) = Pe\cdot S(x_1) + \frac{4+Pe\cdot(x-W)}{x-W}\cdot\frac{1-\tau_1^-}{1+\tau_1^-}\cdot S(x_1^-) + \frac{4}{x-W}\cdot\frac{\gamma_1^- -1}{\gamma_1^- +1}\cdot S(x_1^+),$$

$$\tau_1^- = 2/(2+\sigma^-),\ \gamma_1^- = (2+\theta^-)/2,\ \sigma^- = Pe\,(x_1-W),\ \theta^- = Pe\,(x-x_1).$$

Similarly, we write a scheme of type (2.33) for the segments $[x,\ W]$ and $[x_1,\ x_2]$. Eliminating in the three systems of equations obtained $U^3_{x_1}$ and $U^3_{x_2}$, we obtain a scheme with movable nodes:

$$a_P^7 U^7 = a_E^7 U_E^7 + a_W^7 U_W^7 + F^7(x), \tag{2.34}$$

where

$$a_E^7 = \frac{2^5(1-\gamma_2)}{(E-W)(E-x)(1-\gamma_2^4)},\quad a_W^7 = \frac{4Pe(1-\tau_2)}{(x-W)(1-\tau_2^4)} + \frac{2^5(1-\tau_2)}{(E-W)(x-W)(1-\tau_2^4)},$$

$$a_P^7 = a_W^7 + a_E^7.\quad \tau_2 = 4/(4+\sigma),\ \gamma_2 = (4+\theta)/4$$

$$F^7(x) = Pe\cdot S(x) + \frac{8 + Pe\cdot(x-W)}{x-W}\cdot\frac{(1-\tau_2)^2}{1-\tau_2^4}\cdot\sum_{j=1}^{3}\sum_{i=1}^{j}\tau_2^{i-1}S\left(W + j\frac{x-W}{4}\right)$$

$$-\frac{8}{E-W}\cdot\frac{(1-\gamma_2)^2}{1-\gamma_2^4}\cdot\sum_{j=1}^{3}\sum_{i=1}^{j}\gamma_2^{i-1}S\left(x + (4-j)\frac{E-x}{4}\right).$$

Continuing in this way, we can obtain a scheme with movable nodes $2^k - 1$:

$$a_P^{(2^k-1)} U^{(2^k-1)} = a_E^{(2^k-1)} U_E^{(2^k-1)} + a_W^{(2^k-1)} U_W^{(2^k-1)} + F^{(2^k-1)}(x),$$

where

$$a_E^{(2^k-1)} = \frac{2^{2k+1}(1-\gamma_k)}{(E-W)(E-x)(1-\gamma_k^{2k})},\quad a_W^{(2^k-1)} = \frac{2^{2k+1}Pe(1-\tau_k)}{(x-W)(1-\tau_k^{2k})} + \frac{2^{2k+1}(1-\tau_k)}{(E-W)(x-W)(1-\tau_k^{2k})},$$

$$a_P^{(2^{\Pi}-1)} = a_W^{(2^{\Pi}-1)} + a_E^{(2^{\Pi}-1)}.\ \tau_k = 2^k/(2^k+\sigma),\ \gamma_k = (2^k+\theta)/2^k,$$

$$F^{(2^k-1)}(x) = Pe\cdot S(x) + \frac{2^{k+1} + Pe\cdot(E-W)}{E-W}\frac{(1-\tau_k)^2}{1-\tau_k^{2k}}\sum_{j=1}^{2^k-1}\sum_{i=1}^{j}\tau_k^{i-1}\cdot S\left(x + j\frac{x-W}{2^k}\right)$$

$$-\frac{2^{k+1}}{E-W}\frac{(1-\gamma_k)^2}{1-\gamma_k^{2k}}\sum_{j=1}^{2^k-1}\sum_{i=1}^{j}\gamma_k^{i-1}\cdot S\left(x + (2^k-j)\frac{E-x}{2^k}\right).$$

Figures 2.9 and 2.10 show graphs of approximate solutions of eq. (2.24), obtained by eq. (2.35) for $W=0,\ E=1$, with different moving nodes.

The approximate solutions presented in Figs. 2.9 and 2.10 correspond to different numbers of the moving nodes. Dotted – for $k=1$; dashed – $k=2$; dotted-dotted – $k=3$; long-dashed – $k=4$; rarely dashed – $k=5$; and solid line – exact solution.

It is evident from the graphs that approximate solutions give good results as the number of moving nodes increases.

Note: When obtaining multipoint moving nodes, we proceeded from the upstream scheme. It was possible to proceed from other three-point schemes.

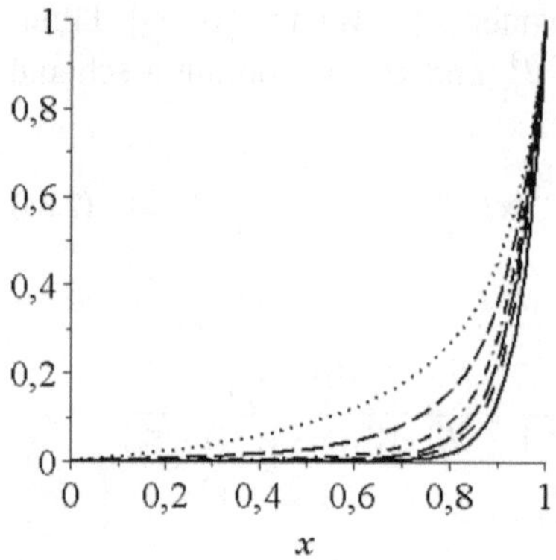

Fig. 2.9: $Pe = 20$, $\Phi_W = 0$, $\Phi_E = 1$, $S(x) = 0$.

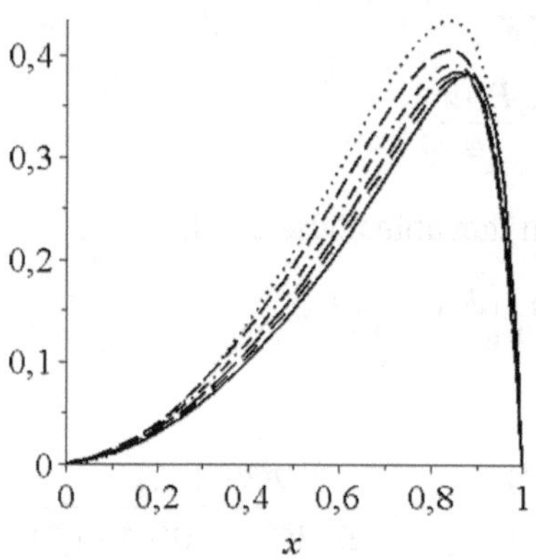

Fig. 2.10: $Pe = 20$, $\Phi_W = 0$, $\Phi_E = 0$, $S(x) = x$.

2.3.2 Analytical method of control volume for one-dimensional convection-diffusion problem

It is known that DEs are obtained on the basis of the integral conservation law. Therefore, the discretization of equations can be carried out using the approximation of integral conservation laws. This method is called the finite volume method [25]. Another name for the method is integral-interpolation [27].

Let us consider a one-dimensional convection-diffusion equation on a finite interval with boundary conditions in the form [25]:

$$\frac{d}{dx}(\rho u \Phi) = \frac{d}{dx}\left(\Gamma \frac{d\Phi}{dx}\right) + S(x), \tag{2.35}$$

$$\Phi(W) = \Phi_W, \quad \Phi(E) = \Phi_E, \tag{2.36}$$

where u is the flow velocity in the x direction, ρ the flow density, Γ is the diffusion coefficient, and $S(x)$ Φ is a given function (source) and is an unknown function. It follows from the continuity equation that $F = \rho u = \text{const}$.

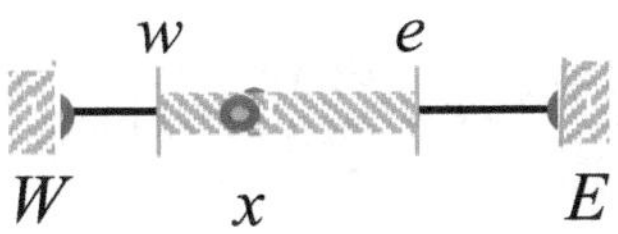

Fig. 2.11: The control volume.

Let us consider eq. (2.35) into segments $[W, E]$. To obtain an approximate analytical solution to the problem using the control volume method, we take an arbitrary point $x \in [W, E]$ and a control volume $[w, e]$ (Fig. 2.11). We assume that the face w is located midway between points W and x, and the face e is midway between points x and E. Integrating eq. (2.35) over the control volume and replacing the derivatives with an upstream scheme, we obtain a zero approximation:

$$(a_E + a_W)\Phi^0 = a_E\Phi_E^0 + a_W\Phi_W^0 + \frac{E-W}{2} . S(x). \tag{2.37}$$

Here $a_E = \frac{\Gamma_e}{E-x} + \max(-F_e, 0); a_W = \frac{\Gamma_w}{x-W} + \max(F_w, 0)$ Since $x \in [W, E]$ an arbitrary point from eq. (2.37) we can determine and obtain an approximate analytical solution to eq. (2.35) under condition (2.36).

Note that, from eq. (2.33), it follows that in the absence of a source $(S(x) \equiv 0)$ on the segments $[W, E]$, the function Φ^0 is monotone.

To improve the approximate solution, we take additional nodes: $x_1 = \frac{x+W}{2}$, $x_2 = \frac{x+E}{2}$. We write an upstream scheme of type (2.33) for the segments $[W, x]$, $[x_1, x_2]$, and $[x, E]$. We obtain a system of three equations. We exclude the obtained systems $\Phi^1(x_1)$ and $\Phi^1(x_2)$, and as a result, we obtain an improved scheme:

$$\begin{gathered}\left[\frac{\beta_1^+}{1+\tau_1} + \frac{\alpha_1^-}{1+\gamma_1}\right]\Phi^1 = \frac{\beta_1^+}{1+\tau_1}\Phi_W^1 + \frac{\alpha_1^-}{1+\gamma_1}\Phi_E^1 + \frac{E-W}{4} \cdot S(x)\\ + \frac{1}{1+\tau_1} \cdot \frac{x-W}{2} \cdot S\left(W + \frac{x-W}{2}\right) + \frac{1}{1+\gamma_1} \cdot \frac{E-x}{2} \cdot S\left(x + \frac{E-x}{2}\right),\end{gathered} \tag{2.38}$$

where

$$\tau_1 = \frac{\beta_1^-}{\beta_1^+}, \ \gamma_1 = \frac{\alpha_1^+}{\alpha_1^-}, \ \beta_1^- = 2D_W + F^-, \ \beta_1^+ = 2D_W + F^+, \ \alpha_1^- = 2D_E + F^-,$$

$$\alpha_1^+ = 2D_E + F^+, D_E = \Gamma/(E-x), \ D_W = \Gamma/(x-W), \ F^- = \max(-F, 0),$$

$$F^+ = \max(F, 0).$$

In eq. (2.38) Φ^1 is the improved value of the unknown function at the nodal point x $(\Phi_W^1 \equiv \Phi_W, \ \Phi_E^1 \equiv \Phi_E)$.

Solving eq. (2.38) with respect to Φ^1 we obtain an improved analytical solution. Again, to improve the solution, we proceed similarly: we write scheme (2.38) for the segment $[W, x]$, $[x_1, x_2]$, and $[x, E]$ and exclude the unknowns at the points x_1 and x_2. Continuing this process, we eventually obtain

$$\begin{aligned}\left[\frac{(1-\tau_k)\beta_k^+}{1-\tau_k^{2k}}+\frac{(1-\gamma_k)\alpha_k^-}{1-\gamma_k^{2k}}\right]\Phi^k &= \frac{(1-\tau_k)\beta_k^+}{1-\tau_k^{2k}}\Phi_W^k+\frac{(1-\gamma_k)\alpha_k^-}{1-\gamma_k^{2k}}\Phi_E^k+\frac{E-W}{2^{k+1}}\cdot S(x)\\ &+\frac{1-\tau_k}{1-\tau_k^{2k}}\cdot\frac{x-W}{2^k}\cdot\sum_{j=1}^{2^k-1}\sum_{i=1}^{j}\tau_k^{i-1}S\left(W+j\frac{x-W}{2^k}\right)\\ &+\frac{1-\gamma_k}{1-\gamma_k^{2k}}\cdot\frac{E-x}{2^k}\cdot\sum_{j=1}^{2^k-1}\sum_{i=1}^{j}\gamma_k^{i-1}S\left(x+(2^k-j)\frac{E-x}{2}\right).\end{aligned}\tag{2.39}$$

Here

$$\tau_k=\frac{\beta_k^-}{\beta_k^+},\ \gamma_k=\frac{\alpha_k^+}{\alpha_k^-},\ \beta_k^-=2^kD_W+F^-,\ \beta_k^+=2^kD_W+F^+,\ \alpha_k^-=2^kD_E+F^-,$$

$$\alpha_k^+=2^kD_E+F^+.$$

In eq. (2.39) Φ^k is the improved value of the unknown function at the nodal point x $(\Phi_W^k\equiv\Phi_W,\ \Phi_E^k\equiv\Phi_E)$. Solving eq. (2.35) with respect to Φ^k we obtain an approximate analytical solution to the original problem.

Examples

Figure 2.12 shows solutions to problems (2.35) and (2.36) with parameters $\Gamma=\text{const}$, $R=\rho u/=20$, $S(x)=0$ for segments [0;1] with boundary conditions. Figure 2.13 shows solutions to problems (2.35) and (2.36) with parameters $R=50$, $S(x)=5\cos 4x$ for segments [0;1] with boundary conditions $\Phi_W=0$, $\Phi_E=1$. The graph shows that as the number of moving nodes increases, the approximate solutions approach the exact ones.

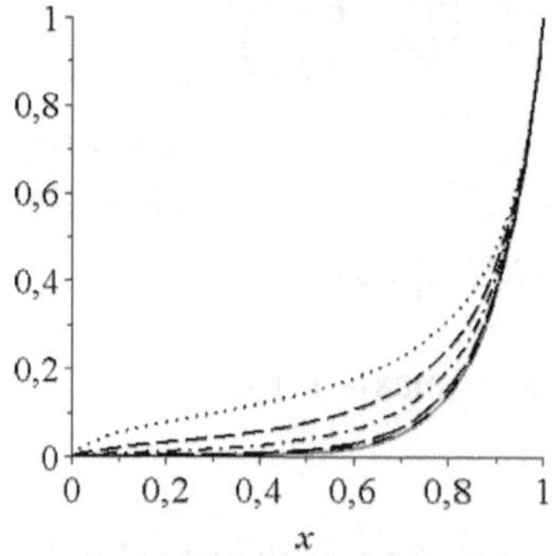

Fig. 2.12: Approximate solutions to problems (2.35) and (2.36): $S(x)=0$.

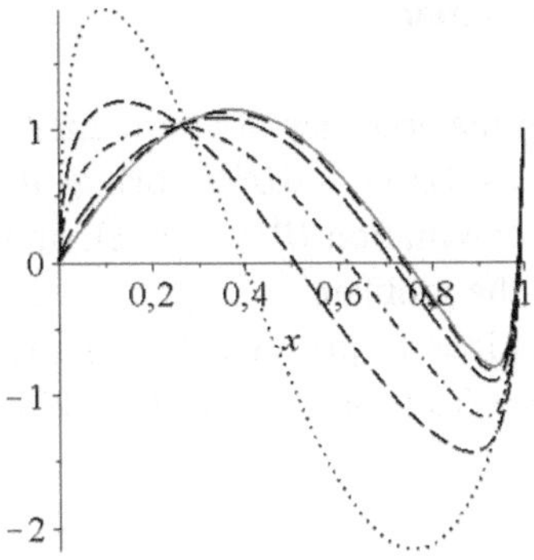

Fig. 2.13: Approximate solutions to problems (2.35) and (2.36): $S(x) = 5\cos 4x$.

In Figs. 2.12 and 2.13, the solid lines are the exact solutions to the problems, the dotted lines are $k = 0$, the dashed lines are $k = 1$, the dotted and dotted lines are $k = 2$, the long-dotted lines are $k = 4$, and the long dotted lines are $k = 6$.

The graphs show that, starting with $k = 6$, the exact and approximate solutions visually coincide.

It is interesting to compare the analytical solution obtained by the finite-difference method (2.30) and the control volume method ($R = 10$, $S(x) = 0$).

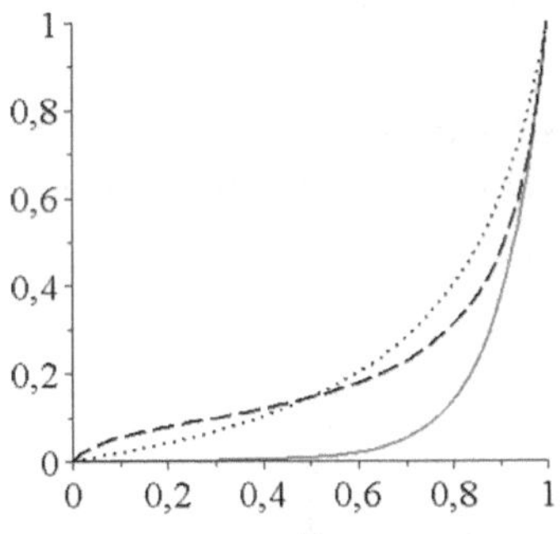

Fig. 2.14: Approximate solution for $k = 0$.

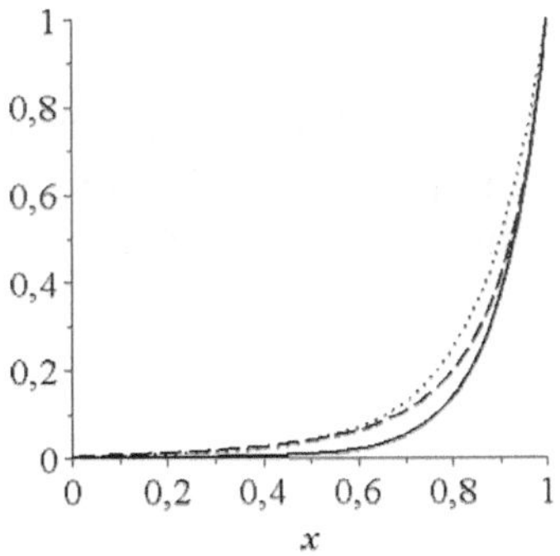

Fig. 2.15: Approximate solution for $k = 1$.

The solid curves in Figs. 2.14 and 2.15 are exact solutions, the dotted ones are the finite-difference method, and the dashed ones are the control volume method.

From Figs. 2.14 and 2.15, it is clear that the solution obtained by the control volume method is preferable.

2.3.3 Increasing the accuracy using Richardson extrapolation

Using the method described in eq. [22], we can increase the accuracy of approximate solutions to the problem. A linear combination $Q^3(x) = -\frac{1}{3}U^1(x) + \frac{4}{3}U^3(x)$ brings us closer to the solution more accurately. With a linear combination $U^1(x)$, $U^3(x)$, and $U^7(x)$ in the form, we obtain a more refined solution to the problem.

Figure 2.16 shows the graphs of approximate solutions to problems (2.35) and (2.36) obtained using Richardson extrapolation at $W = 0$. The solid line in Figs. 2.18 and 2.19 is the exact solution.

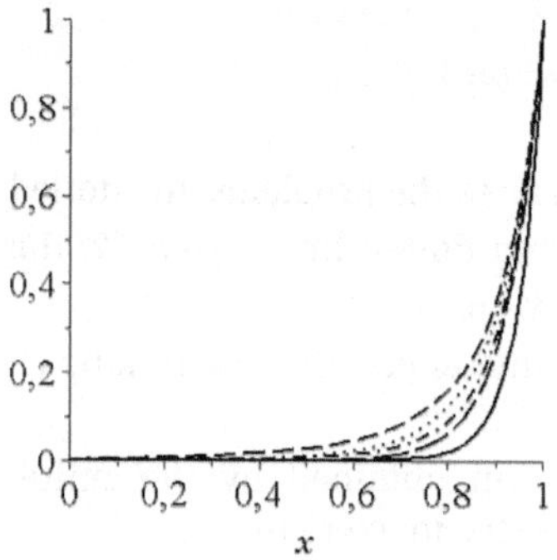

Fig. 2.16: $\Phi_W = 0$, $\Phi_E = 1$, $S(x) = 0$.

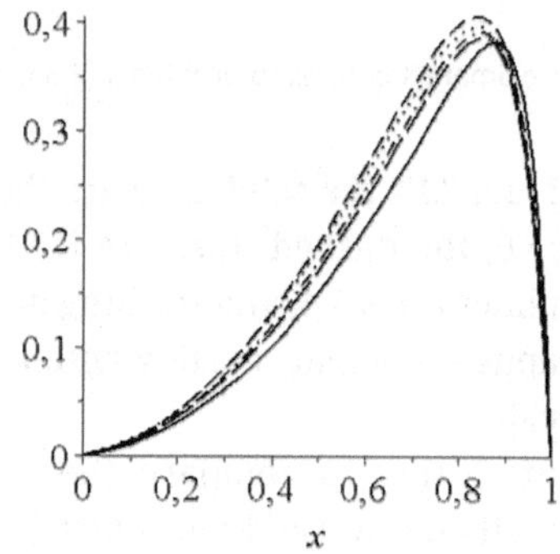

Fig. 2.17: $\Phi_W = 0$, $\Phi_E = 0$, $S(x) = x$.

Figures 2.16 and 2.17 show comparisons of the results obtained on the basis of Richardson extrapolation (dashed line – $U^3(x)$, dotted – $Q^3(x)$, dotted-dotted – $U^7(x)$, and long dotted – $Q^7(x)$ ($Pe = 20.$)).

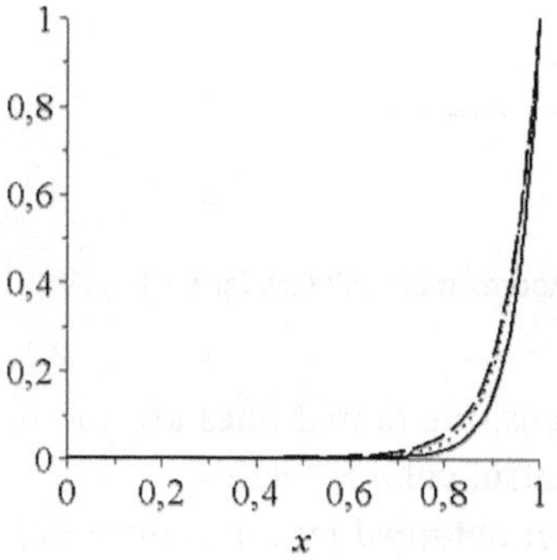

Fig. 2.18: $\Phi_W = 0$, $\Phi_E = 1$, $S(x) = 0$.

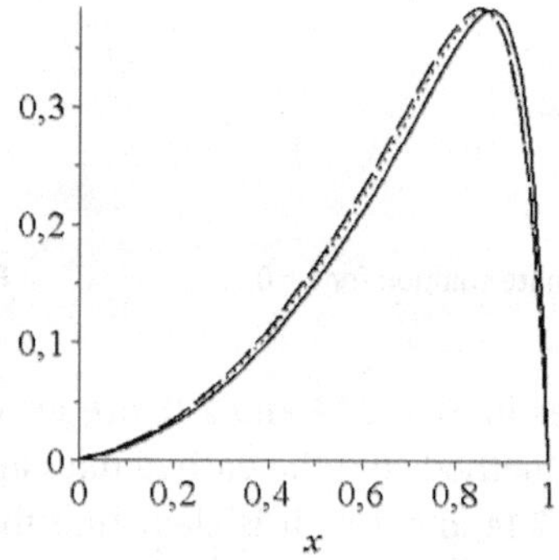

Fig. 2.19: $\Phi_W = 0$, $\Phi_E = 0$, $S(x) = x$.

Figures 2.18 and 2.19 also provide comparisons $U^{15}(x)$, $Q^{15}(x)$ with exact solution $(Pe = 20)$.

Figures 2.16–2.19 allow us to state that Richardson extrapolation allows us to obtain a more accurate solution to the problem.

The following sections present some applications of the MNM. Using multipoint moving nodes, improved schemes are constructed for the convective-diffusion problem. An instruction for schemes using the Richardson extrapolation method is proposed. Some properties of the schemes are also presented for study using the moving nodes.

2.4 Obtaining discrete compact schemes for the convective-diffusion problem of the MNM

The construction and analysis of the discretization of a singularly perturbed ODE of the second order is of great interest. The convection-diffusion equation is the basic one in modeling fluid flow at high Reynolds numbers and in convective mass transfer at high Peclet numbers. Many works are devoted to this topic [8, 11–14, 16, 17, 19, 32, 33].

Numerical solutions of the convection-diffusion equation often exhibit numerical oscillations. In practical calculations, many authors have observed parasitic oscillations at high Peclet numbers when using a central approximation for the convective term [25]. On the other hand, the upwind scheme usually leads to unpleasant artificial numerical diffusion [25, 34, 35].

This dilemma is central to numerical solutions of convection-diffusion problems.

To compute approximate solutions of a partial DE, some form of local approximation must be used. This means that the solution values at each node are used to generate an approximate solution value. For finite differences, one usually tries to make the local domain as compact as possible, for example, using only neighboring nodes when updating at a node.

If we consider the approximation of the convection-diffusion problem on a uniform grid, we can observe that most of the literature is concerned with the choice between the scheme in the three-point pattern: (W,P,E). To obtain a high-order approximation, the number of points in the computational stencil has to be increased [17].

In this section, we apply the structure outlined in the previous paragraphs to develop a new finite-difference scheme. However, it is important to note that this procedure is not easily generalizable to partial differential equations with variable coefficients.

2.4.1 Construction of compact schemes of the convection-diffusion problem based on the FDM

We consider the convection-diffusion DE:

$$\frac{d\Phi}{dx} = \frac{1}{Pe}\frac{d^2\Phi}{dx^2} + S(x), \tag{2.40}$$

with the boundary conditions:

$$\Phi(0) = \Phi_0,\ \Phi(1) = \Phi_1, \tag{2.41}$$

where Pe is the Peclet number, $(Pe = \rho v L/\Gamma)$ v is the velocity, ρ is density, L is the length scale, Γ is the diffusion coefficient, x is the dimensionless coordinate, and $S(x)$ is the source.

We introduce a nonuniform grid on [0,1]:

$$\Omega = \{x_i,\ i = 0, 1, 2, \ldots, N,\ 0 = x_0 < x_1 < \cdots < x_{i-1} < x_i < x_{i+1} < \cdots < x_N = 1\}.$$

In Section 2.4.1, using movable nodes, we constructed an analytical solution to problems (2.40) and (2.41). When constructing compact schemes, we rely on the upstream scheme, which is monotone for any Peclet numbers.

We rewrite the upstream scheme (2.25) for the segment (W,E):

$$Pe\frac{U^1 - U^1_W}{x - W} = \frac{2}{(E - W)}\left(\frac{U^1_E - U^1}{E - x} - \frac{U^1 - U^1_W}{x - W}\right) + Pe \cdot S(x). \tag{2.42}$$

In eq. (2.42), the equation connects unknown function at three points: W, x, E, that is, eq. (2.42) is written in a three-point pattern. Now we write eq. (2.42) for an arbitrary internal node x_i, which connect with neighboring nodes x_{i-1}, x_{i+1}. Then

$$Pe\frac{U^1_i - U^1_{i-1}}{x_i - x_{i-1}} = \frac{2}{(x_{i+1} - x_{i-1})}\left(\frac{U^1_{i+1} - U^1_i}{x_{i+1} - x_i} - \frac{U^1_i - U^1_{i-1}}{x_i - x_{i-1}}\right) + Pe \cdot S(x_i). \tag{2.43}$$

Here $i = 1, 2, \ldots, i, \ldots, N - 1$ and U^1_i means the approximate value of the unknown function at node x_i.

This scheme can be rewritten as follows:

$$a^1_P U^1_P = a^1_E U^1_E + a^1_W U^1_W + F^1_i. \tag{2.44}$$

But now

$$a^1_E = \frac{2}{(x_{i+1} - x_{i-1})(x_{i+1} - x_i)},\quad a^1_W = \frac{Pe}{(x_i - x_{i-1})} + \frac{2}{(x_{i+1} - x_{i-1})(x_{i+1} - x_i)},$$
$$a^1_P = a^1_E + a^1_W,\ F^1_i = Pe \cdot S(x_i). \tag{2.45}$$

To increase accuracy, based on three movable nodes (2.33),

$$a_P^3 U_P^3 = a_E^3 U_E^3 + a_W^3 U_W^3 + F_i^3 \tag{2.46}$$

where

$$a_E^3 = \frac{8}{(x_{i+1}-x_{i-1})(x_{i+1}-x_i)(1+\gamma_1)},\ a_W^3 = \frac{2Pe}{(x_i-x_{i-1})(1+\tau_1)} + \frac{8}{(x_{i+1}-x_{i-1})(x_i-x_{i-1})(1+\tau_1)},$$

$$a_P^3 = a_W^3 + a_E^3,\ \theta = Pe\,(x_{i+1}-x_i),\ \sigma = Pe\,(x_i - x_{i-1}),$$

$$\tau_1 = 2/(2+\sigma),\ \gamma_1 = (2+\theta)/2$$

$$F_i^3 = Pe\cdot S(x_i) + \frac{4+Pe\cdot(x_{i+1}-x_{i-1})}{x_{i+1}-x_{i-1}}\cdot\frac{1-\tau_1}{1+\tau_1}\cdot S(x_{i-1/2}) + \frac{4}{x_{i+1}-x_{i-1}}\cdot\frac{\gamma_1-1}{\gamma_1+1}\cdot S(x_{i+1/2})$$

$x_{i-1/2} = 0,5(x_{i-1}+x_i),\ x_{i+1/2} = 0,5(x_i + x_{i+1})$.

Based on with movable nodes 2^k-1 (2.35), we have

$$a_P^{(2^k-1)} U_P^{(2^k-1)} = a_E^{(2^k-1)} U_E^{(2^k-1)} + a_W^{(2^k-1)} U_W^{(2^k-1)} + F_i^{(2^k-1)}, \tag{2.47}$$

where

$$a_E^{(2^k-1)} = \frac{2^{2k+1}(1-\gamma_k)}{(x_{i+1}-x_{i-1})(x_{i+1}-x_i)(1-\gamma_k^{2k})},\ a_W^{(2^k-1)} = \frac{2^{2k+1}Pe(1-\tau_k)}{(x_i-x_{i-1})(1-\tau_k^{2k})}$$

$$+ \frac{2^{2k+1}(1-\tau_k)}{(x_{i+1}-x_{i-1})(x_i-x_{i-1})(1-\tau_k^{2k})},$$

$$a_P^{(2^k-1)} = a_W^{(2^k-1)} + a_E^{(2^k-1)}.\ \tau_k = 2^k/(2^k+\sigma),\ \gamma_k = (2^k+\theta)/2^k$$

$$F_i^{(2^k-1)} = Pe\cdot S(x_i) + \frac{2^{k+1}+Pe\cdot(x_{i+1}-x_{i-1})}{x_{i+1}-x_{i-1}}\frac{(1-\tau_k)^2}{1-\tau_k^{2k}}\sum_{j=1}^{2^k-1}\sum_{i=1}^{j}\tau_k^{i-1}\cdot S\left(x_i + j\frac{x_i-x_{i-1}}{2^k}\right)$$

$$-\frac{2^{k+1}}{x_{i+1}-x_{i-1}}\frac{(1-\gamma_k)^2}{1-\gamma_k^{2k}}\sum_{j=1}^{2^k-1}\sum_{i=1}^{j}\gamma_k^{i-1}\cdot S\left(x_i + (2^k-j)\frac{x_{i+1}-x_i}{2^k}\right).$$

Let us consider numerical experiments.

Figures 2.19 and 2.20 show graphs of the solution of problem (2.40), (2.41) for $Pe = 50$ on segments [0;1] with boundary conditions $\Phi_0 = 0,\ \Phi_1 = 1$. Figure 2.19 corresponds to $S(x) = 5\cos 4x$, and the graphs in Fig. 2.20 are obtained for

$$S(x) = \begin{cases} 10 - 50x & \text{если } \le 0{,}3 \\ 50x - 20 & \text{если } 0{,}3 < x < 0{,}4 \\ 0 & \text{если } 0{,}4 < x \end{cases} \tag{2.48}$$

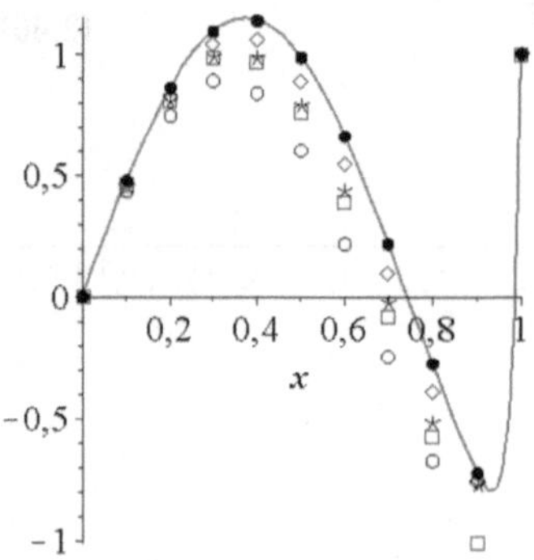

Fig. 2.20: $Pe = 50$. $S(x) = 5\cos 4x$..

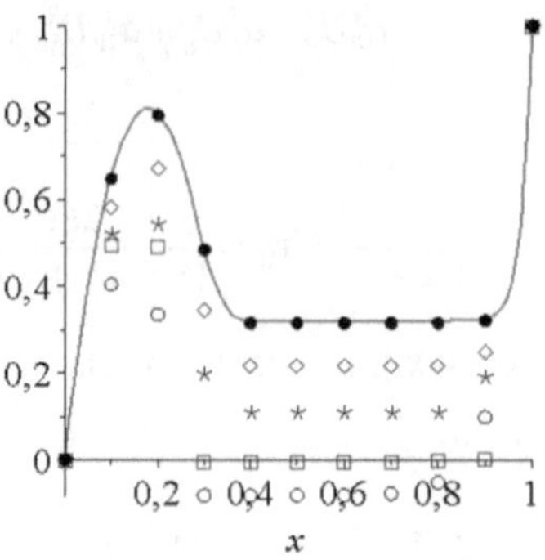

Fig. 2.21: $Pe = 50$. $S(x)$ (2.48).

The numerical results are obtained for $h = 0.1$, and the grid Peclet number is equal to 5. The solid lines are the graphs of the exact solutions of the problem. The circle symbols are obtained for the upwind scheme, the rectangles are according to the Patankar scheme, the stars are according to eq. (2.46), the diamond is according to eq. (2.47) for $k = 2$, and the circles are according to eq. (2.47) for $k = 7$.

Table 2.1 shows the root-mean-square errors $\sigma = \sqrt{\sum_1^N ((\Phi x_i) - U_i)^2/N}$ for according to the considered schemes. Φ_i is the exact solution at nodal points, U_i is the numerical solution obtained according to the considered schemes, and N is the number of nodes.

Tab. 2.1: The root-mean-square error.

Scheme	Against the flow	Patankar's	(2.2.7)	(2.2.8), $k = 2$	(2.2.8), $k = 7$
$S(x) = 5\cos 4x$	0.361	0.300	0.192	0.096	0.004
$S(x)$ по (2.9)	0.282	0.199	0.154	0.077	0.003

From Figs. 2.19 and 2.20 and from Tab. 2.1 it is clear that the proposed schemes give good results.

2.4.2 Construction of compact schemes of convection-diffusion problem based on finite volume method

Finite volume method is one of the methods that can give good approximate solution of the problem [25]. Here we investigate application of finite volume method for solving convection-diffusion equation for construction of compact schemes [3].

Approximation of convective term in convection-diffusion problems is a central problem. When convective term of the equation is approximated by central difference, which leads to second-order schemes that are stable only for sufficiently small cell size. Various modifications of upwind schemes have been proposed to increase the accuracy of constructed schemes [8, 11–14, 16, 17, 19, 32, 33]. We construct a number of modifications of upstream strategies that provide increased accuracy.

The basic strategy of all finite volume methods is to write the DE in conservative form, integrate it into small regions (called "cells" or "finite volumes"), and transform each such integral over the cell boundary.

Our goal is to construct a high-quality scheme for the problem:

$$\frac{d}{dx}(\rho u\Phi)=\frac{d}{dx}\left(\Gamma\frac{d\Phi}{dx}\right)+S(x), \tag{2.49}$$

$$\Phi(0)=\Phi_0,\quad \Phi(1)=\Phi_1 \tag{2.50}$$

based on the control volume method. The technique for obtaining the scheme is similar to that described in Section 2.5.1.

We introduce a nonuniform grid on [0,1]:

$$\Omega=\{x_i,\ i=0,1,2,\ldots,N,\ 0=x_0<_1<\cdots<x_{i-1}<x_i<x_{i+1}<\cdots<x_N=1\}.$$

Using moving nodes, an analytical solution of problem (2.49), eq. (2.50) was constructed by the control volume method in the form:

$$\begin{aligned}\left[\frac{(1-\tau_k)\beta_k^+}{1-\tau_k^{2^k}}+\frac{(1-\gamma_k)\alpha_k^-}{1-\gamma_k^{2^k}}\right]U^k&=\frac{(1-\tau_k)\beta_k^+}{1-\tau_k^{2^k}}U_W^k+\frac{(1-\gamma_k)\alpha_k^-}{1-\gamma_k^{2^k}}U_E^k+\frac{E-W}{2^{k+1}}\cdot S(x)\\&+\frac{1-\tau_k}{1-\tau_k^{2^k}}\cdot\frac{x-W}{2^k}\cdot\sum_{j=1}^{2^k-1}\sum_{i=1}^{j}\tau_k^{i-1}s\left(W+j\frac{x-W}{2^k}\right)\\&+\frac{1-\gamma_k}{1-\gamma_k^{2^k}}\cdot\frac{E-x}{2^k}\cdot\sum_{j=1}^{2^k-1}\sum_{i=1}^{j}\gamma_k^{i-1}s\left(x+(2^k-j)\frac{E-x}{2}\right).\end{aligned} \tag{2.51}$$

Here

$$\tau_k=\frac{\beta_k^-}{\beta_k^+},\ \gamma_k=\frac{\alpha_k^+}{\alpha_k^-},\ \beta_k^-=2^kD_W+F^-,\ \beta_k^+=2^kD_W+F^+,\ \alpha_k^-=2^kD_E+F^-,$$

$$\alpha_k^+=2^kD_E+F^+,\ F^-=\max(-F,0),F^+=\max(F,0),$$

$$D_E=\Gamma/(E-x),\ D_W=\Gamma/(x-W).$$

Now we write eq. (2.51) for an arbitrary internal node x_i, which connects with neighboring nodes x_{i-1}, x_{i+1}. Then

$$\left[\frac{(1-\tau_k)\beta_k^+}{1-\tau_k^{2k}}+\frac{(1-\gamma_k)\alpha_k^-}{1-\gamma_k^{2k}}\right]U_P{}^k=\frac{(1-\tau_k)\beta_k^+}{1-\tau_k^{2k}}U_W^k+\frac{(1-\gamma_k)\alpha_k^-}{1-\gamma_k^{2k}}U_E^k+\frac{x_{i+1}-x_{i-1}}{2^{k+1}}\cdot S(x_i)$$
$$+\frac{1-\tau_k}{1-\tau_k^{2k}}\cdot\frac{x_i-x_{i-1}}{2^k}\cdot\sum_{j=1}^{2^k-1}\sum_{m=1}^{j}\tau_k^{m-1}S\left(x_{i-1}+j\frac{x_i-x_{i-1}}{2^k}\right)$$
$$+\frac{1-\gamma_k}{1-\gamma_k^{2k}}\cdot\frac{x_{i+1}-x_i}{2^k}\cdot\sum_{j=1}^{2^k-1}\sum_{m=1}^{j}\gamma_k^{m-1}S\left(x_i+(2^k-j)\frac{x_{i+1}-x_i}{2}\right).$$

Here $D_E=\Gamma/(x_{i+1}-x_i)$ and $D_W=\Gamma/(x_i-x_{i-1})$.

Equation (2.51) is a compact scheme for problems (2.49) and (2.50) constructed on the basis of the control volume method taking into account the MNM.

2.4.3 Construction of schemes using the Richardson extrapolation method

The Richardson method is used to solve grid problems on a sequence of grids. The method consists of performing calculations for the same scheme, with different steps. Then we have several grid solutions. Based on the grid solutions, a certain linear combination is compiled. The resulting linear combination has a higher order of accuracy [6].

The theory and applications of this method are presented in [22].

Creation of new schemes using Richardson extrapolation based on the schemes given in Section 2.4.3.

The accuracy of scheme (2.43), with uniform placement of grid nodes, is $O(h)$. Scheme (2.46) has the order of $O(h/2)$. For a linear combination $Q^3(x_i)=-\frac{1}{3}U^1(x_i)+\frac{4}{3}U^3(x_i)$ we obtain an approximation error $O(h^2)$ on a uniform grid. A linear combination $U^1(x_i)$, $U^3(x_i)$, and $U^7(x_i)$ in the form $Q^7(x_i)=\frac{1}{45}U^1(x_i)-\frac{4}{9}U^3(x_i)+\frac{64}{45}U^7(x_i)$ has an approximation order of $O(h^4)$. Let us consider $N=10$, $S(x)=x^2$, and $Pe=30$. Table 2.2 shows the absolute difference between the exact and approximate solutions according to the schemes.

Tab. 2.2: Absolute difference between the exact and approximate solutions.

x_x	0.1	0.2	0.3	0.4	0.5	0.6	0.7	0.8	0.9
Scheme	0.001	0.004	0.007	0.011	0.017	0.025	0.039	0.073	0.160
$U^3(x_i)$	0.001	0.002	0.004	0.006	0.008	0.012	0.018	0.034	0.089
$U^7(x_i)$	0.000	0.001	0.002	0.003	0.005	0.007	0.010	0.019	0.046
$Q^3(x_i)$	0.000	0.001	0.002	0.004	0.006	0.007	0.010	0.021	0.065
$Q^7(x_i)$	0.000	0.001	0.001	0.002	0.003	0.005	0.007	0.014	0.030

Table 2.3 shows the root-mean-square error $\sigma = \sqrt{\sum_1^N (\Phi(x_i) - U_i)^2/N}$ for the considered schemes. (Φx_i) is the exact solution at nodal points and U_i is the numerical solution obtained for the considered schemes.

Tab. 2.3: Root-mean-square error.

Scheme	(3)	$U^3(x)$	$U^7(x)$	$Q^3(x)$	$Q^7(x)$
$S = x^2$, $Pe = 50$, $\Phi_W = 0$, $\Phi_E = 1$	0.047	0.023	0.011	0.015	0.006
$S = 10$, $Pe = 50$, $\Phi_W = 0$, $\Phi_E = 1$	0.033	0.017	0.008	0.011	0.005
$S = x^2$, $Pe = 100$, $\Phi_W = 0$, $\Phi_E = 1$	0.034	0.014	0.006	0.008	0.003
$S = 5\cos(4\pi x)$, $Pe = 50$, $\Phi_W = 0$, $\Phi_E = 1$.	0.213	0.120	0.061	0.090	0.038

Figures 2.22 and 2.23 show the numerical solutions for $\Phi_W = 0$ and $\Phi_E = 0$.

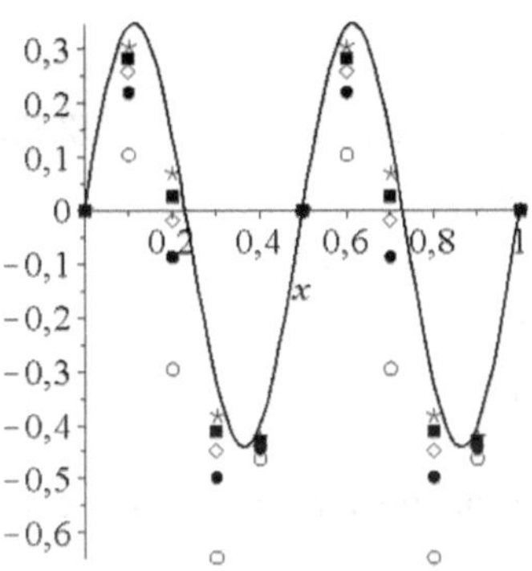

Fig. 2.22: Comparison of schemes for $Pe = 100$ and $S = 5\cos 4\pi x$.

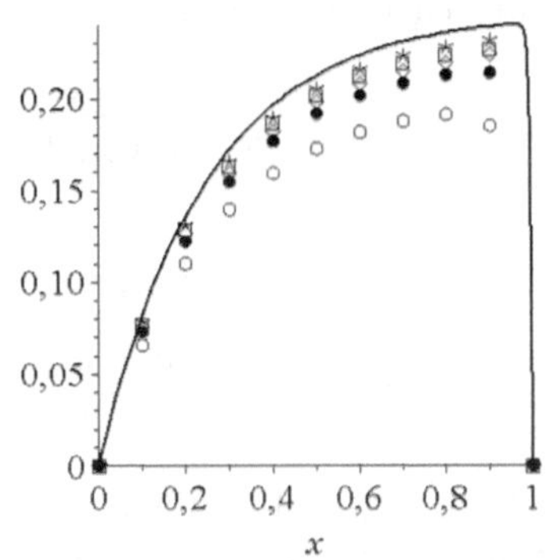

Fig. 2.23: Comparison of schemes for $Pe = 200$ and $S = \exp(-4x)$.

In Figs. 2.22 and 2.23, solid curve is the exact solution, circle obtained by scheme U^1, circle by U^3, solid rectangle by U^7, diamond by Q^3, and star by Q^7.

From the graphs in Figs. 2.22 and 2.23 and from Tabs 2.2 and 2.3, it is clear that the linear combination according to Richardson allows to obtain a more improved scheme.

2.5 Effect of the choice of the profile on the edge of the control volume on the quality of difference schemes

When obtaining discrete analogs of convective-diffusion problems, given above, based on multipoint moving nodes, it was possible to construct higher-quality compact schemes in a three-point template. However, there is another approach to improving the quality of the scheme based on the choice of the solution profile.

Starting with the work of Leonard [17], in order to improve the results of the numerical solution, attempts were made to improve the algorithm, which is constructed in a five-point template.

In all of the listed schemes (except for the upstream scheme), the conditions of boundedness and non-negativity of the coefficients are violated.

Here we propose to improve the scheme based on the choice of the solution profile on the edge of the control volume in the three-point template of the convective-diffusion problem. The upwind scheme with one-sided differences is taken as the initial scheme. In the QUICK scheme [19], quadratic upwind interpolation is used to determine the convective flow. Here we use the solution obtained by the upwind scheme based on the moving node method.

For simple cases, the MNM allows us to obtain an analytical representation of the solution between the nodal points of the boundary value problem. Based on this representation, we can construct a higher-quality discrete scheme.

Integrating eq. (2.40) over the control volume

$$\Phi_e - \Phi_w = \frac{1}{Pe}\left(\frac{d\Phi}{dx}\right)_e - \frac{1}{Pe}\left(\frac{d\Phi}{dx}\right)_w + \int_w^e S(x)dx. \tag{2.52}$$

Replacing in eq. (2.52) the derivatives with difference relations, we have

$$\Phi_e - \Phi_w = \frac{1}{Pe}\frac{\Phi_E - \Phi_P}{x_E - x_P} - \frac{1}{Pe}\frac{\Phi_P - \Phi_W}{x_P - x_W} + (x_e - x_w)f_P. \tag{2.53}$$

Here $f_P = \frac{1}{x_e - x_w}\int_w^e S(x)dx$. Depending on the type of the function Φ profile on the control volume, different schemes are obtained.

Let the profile Φ be piecewise constant in each control volume. Then assuming $\Phi_e = \Phi_P$ and $\Phi_w = \Phi_W$, we have a upwind scheme:

$$\Phi_P - \Phi_W = \frac{1}{Pe}\frac{\Phi_E - \Phi_P}{x_E - x_P} - \frac{1}{Pe}\frac{\Phi_P - \Phi_W}{x_P - x_W} + (x_e - x_w)f_P. \tag{2.54}$$

If the profile Φ is linear between the nodes and the edges of the control volume are located in the middle between the nodal points, we have a scheme with central differences:

$$\frac{\Phi_E + \Phi_P}{2} - \frac{\Phi_P + \Phi_W}{2} = \frac{1}{Pe}\frac{\Phi_E - \Phi_P}{x_E - x_P} - \frac{1}{Pe}\frac{\Phi_P - \Phi_W}{x_P - x_W} + (x_e - x_w)f_P. \tag{2.55}$$

To improve the accuracy of the schemes, many authors have recommended various schemes [12, 13, 27]. All of these schemes are multipoint (more than three). Here we propose a method for improving three-point schemes.

From eq. (2.15) we obtain:

$$\begin{aligned}\Phi_P &= \frac{x_P - x_W}{Pe(x_E - x_P)(x_P - x_W) + x_E - x_W}\Phi_E + \frac{(x_E - x_P)(1 + Pe(x_P - x_W))}{Pe(x_E - x_P)(x_P - x_W) + x_E - x_W}\Phi_W + \\ &= \frac{(x_P - x_W)\Phi_E + (x_E - x_P)(1 + Pe(x_P - x_W))\Phi_W}{Pe(x_E - x_P)(x_P - x_W) + x_E - x_W}\end{aligned} \tag{2.56}$$

If the nodes x_E and x_W are fixed, and the node x_P is movable, we obtain a profile Φ_P between the nodes $x_E$and x_W. We use this profile in eq. (2.53) to determine Φ_e and Φ_w.

To improve scheme (2.55), we proceed as follows. Equation (2.56) connects Φ at three nodes (x_W, x_P, x_E), if we apply eq. (2.56) for nodes (x_W, x_w,x_P), we have

$$\Phi_w = \frac{2 + R_h}{4 + R_h}\Phi_P + \frac{2}{4 + R_h}\Phi_E + \frac{R_h}{2(R_h + 4)}\frac{h}{4}f_w. \tag{2.57}$$

Similarly, for nodes (x_P, x_e,x_E), we have

$$\Phi_e = \frac{2 + R_h}{4 + R_h}\Phi_P + \frac{2}{4 + R_h}\Phi_E + \frac{R_h\, h}{2(R_h + 4)}f_e. \tag{2.58}$$

Substituting eqs. (2.57) and (2.58) into eq. (2.53), we have

$$\left[\frac{R_h^2}{4 + R_h} + 2\right]\Phi_P = \left[1 + \frac{2 + R_h}{4 + R_h}\right]\Phi_W + \left[1 - \frac{2R_h}{4 + R_h}\right]\Phi_E + h\,R_h f_P - \frac{h \cdot R_h^2}{2(4 + R_h)}(f_e - f_w). \tag{2.59}$$

The conditions $R_h < 4$ ensure the positivity of the coefficients and the stability of scheme (2.59).

Proceeding in the same way as in deriving (2.59), but for the profile we use eq. (2.59); for a uniform step we obtain eq. (2.60):

$$\begin{aligned}\left[\frac{(4 + R_h)^2 - 16}{(4 + R_h)^2 + 16}\right]\Phi_P &= \left[\frac{1}{R_h} - \frac{16}{(4 + R_h)^2 + 16}\right]\Phi_E + \left[\frac{(4 + R_h)^2}{(4 + R_h)^2 + 16} + \frac{1}{R_h}\right]\Phi_W \\ &+ hS(P) + \frac{h(8R_h + 8R_h^2)}{2\left[(4 + R_h)^2 + 16\right]} \cdot (S(x_w) - S(x_e)).\end{aligned} \tag{2.60}$$

Test problems

1. Consider the equation

$$\frac{du}{dx} = \frac{1}{Pe}\frac{d^2u}{dx^2} + \sin \pi x,$$

with boundary conditions $u(0) = u(1) = 0$. Table 2.4 shows the maximum absolute differences of the schemes calculated at the nodal points (u – the exact solution of the problem, u_1 – the solution obtained by the upwind scheme, u_2 – by the power law, u_3 – by the Leonard scheme, u_4 by eq. (2.59), and u_5 – by scheme (2.60).

Tab. 2.4: Maximum absolute differ-ences of the schemes.

Pe	R_h	max $\lvert u-u_1\rvert$	max $\lvert u-u_2\rvert$	max $\lvert u-u_3\rvert$	max $\lvert u-u_4\rvert$	max $\lvert u-u_5\rvert$
100	10	0.0526	0.03770	0.1801	0.03701	0.00077
1,000	100	0.0470	0.0464	0.2732	0.01607	0.00927

2. Consider the equation

$$\frac{du}{dx}=\frac{1}{Pe}\frac{d^2u}{dx^2}+s(x),$$

with boundary conditions $u(0)=0,\ \ u(1)=1$, with a source

$$s(x)=\begin{cases}10-50x, & 0\le x\le 0.3,\\ 50x-20, & 0.3<x\le 0.4,\\ 0, & 0.4<x\le 1.\end{cases}$$

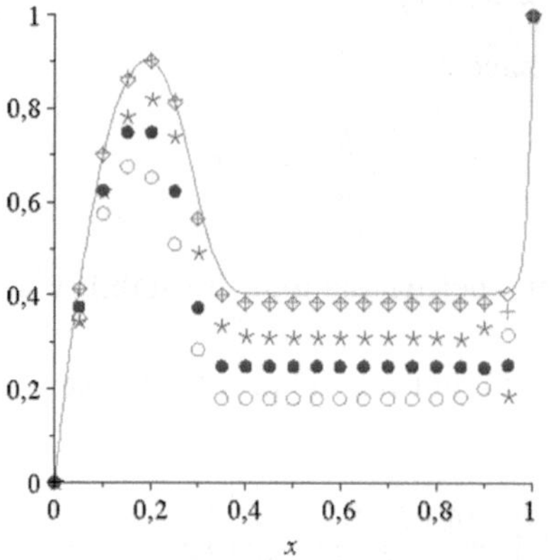

Fig. 2.24: Comparison of different schemes ($Pe=100,\ \ R_h=5$).

In Fig. 2.24, solid line is exact solution, circle is upwind scheme, circle is Patankar scheme, asterisk is Leonard scheme, + is scheme (2.59), and diamond is by (2.60).

Figure 2.6 shows that scheme (2.60) gives the best results. Leonard's scheme gives an incorrect solution near the right boundary. Scheme (2.59) also has a slight non-monotonicity. This is due to the fact that scheme (2.59) is stable at $R_h<4$.

In Fig. 2.25, solid line is the exact solution, circle is the upstream scheme, circle is the Patankar scheme, asterisk is the Leonard scheme, + is scheme (2.59), and diamond by eq (2.60).

Figure 2.25 shows that for large grid Peclet numbers, the upstream and Patankar schemes give close results. Scheme (2.60) gives the best results. This is also evident from Tab. 2.5, which shows a comparison of the considered schemes (SDS is the central scheme).

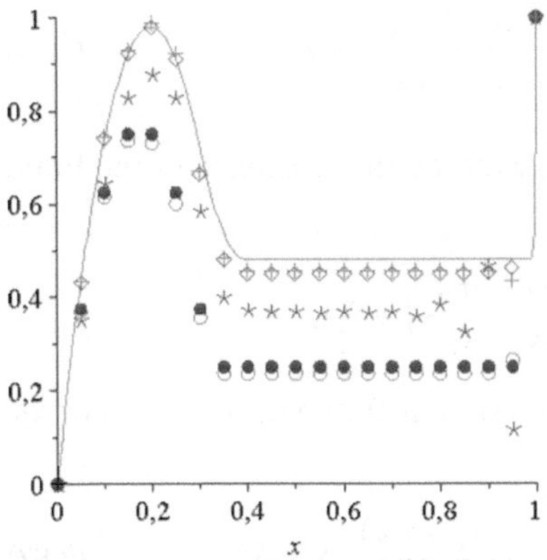

Fig. 2.25: Comparison of different schemes ($Re = 500$, $R_h = 25$). solid line is the exact solution, circle is the upstream scheme, circle is Patankar's scheme, asterisk is Leonard's scheme, + is the scheme (2.59), diamond is by (2.60).

Tab. 2.5: Comparision of the schemes.

Scheme	Re	h	R_h	Max$\lvert u - u_p \rvert$	$\frac{\sum \lvert u_i - (u_p)_i \rvert}{\sum u_i}$
Against the flow	100	1/40	2.5	0.1627	0.2116
	100	1/20	5	0.3258	0.4224
	500	1/20	25	0.3650	0.4101
Power	100	1/40	2.5	0.0833	0.1057
	100	1/20	5	0.2385	0.3025
	500	1/20	25	0.3454	0.3868
(2.8)	100	1/40	2.5	0.0164	0.0169
	100	1/20	5	0.0460	0.0401
	500	1/20	25	0.0531	0.0398
(2.12)	100	1/40	2.5	0.0129	0.00840
	100	1/20	5	0.0452	0.0358
	500	1/20	25	0.0571	0.0404
QUICK	100	1/40	2,5	0.0700	0.0701
	100	1/20	5	0.2231	0.1931
	500	1/20	25	0.3653	0.2055
CDS	100	1/40	2.5	0.1237	0.0062
	100	1/20	5	0.3033	0.0467
	500	1/20	25	0.5136	0.1355

2.6 Improving the circuit using flow equality

The quality of the circuit can be improved using movable nodes. We demonstrate this method based on the counter-flow circuit (2.52) written as

$$\frac{\Phi_P - \Phi_W}{x_P - x_W} = \frac{2}{Pe(x_E - x_W)}\left(\frac{\Phi_E - \Phi_P}{x_E - x_P} - \frac{\Phi - \Phi_W}{x_P - x_W}\right) + S(x_P). \tag{2.61}$$

In eq. (2.61), we pass to the limit $x_E \to x_P$ at and, assuming the existence of the limit, we have

$$\frac{\Phi_P - \Phi_W}{x_P - x_W} = \frac{2}{Pe(x_P - x_W)}\left(\frac{d\Phi_P^-}{dx_P} - \frac{\Phi_P - \Phi_W}{x_P - x_W}\right) + S(x_P).$$

Here $d\Phi_P^-/dx_P$ is the left-hand derivative of the unknown function at the point x_P. Hence

$$\frac{d\Phi_P^-}{dx_P} = \frac{2 + Pe(x_P - x_W)}{2} \cdot \frac{\Phi_P - \Phi_W}{x_P - x_W} - \frac{Pe(x_P - x_W)}{2} \cdot S(x_P). \tag{2.62}$$

Similarly, taking an arbitrary point $x \in (x_P, x_E)$ and passing to the limit $x \to x_P$, we find

$$\frac{d\Phi_P^+}{dx_P} = \frac{2}{2 + Pe(x_E - x_P)} \cdot \frac{\Phi_E - \Phi_P}{x_E - x_P} + \frac{Pe(x_E - x_P)}{2 + Pe(x_E - x_P)} \cdot S(x_P).$$

Equating the flows $d\Phi^+/dx = d\Phi^-/dx$, we obtain an improved circuit:

$$c_P\Phi_P = a_P\Phi_W + b_P\Phi_E + d_PS(x_P), \tag{2.63}$$

where

$$a_P = \frac{2 + Pe(x_P - x_W)}{(x_P - x_W)}, \quad b_P = \frac{2}{[2 + Pe(x_E - x_P)](x_E - x_P)}, \quad c_P = a_P + b_P.$$

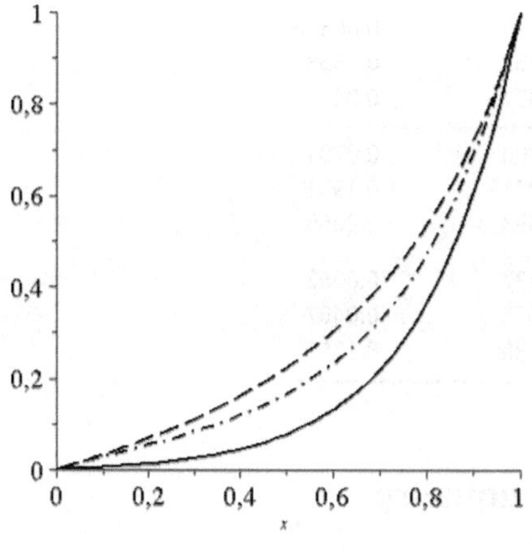

Fig. 2.26: Comparison of circuits. Solid curve – exact solution, dashed by eq. (2.52), and dotted-dotted by eq. (2.63).

Figure 2.25 shows a comparison of the exact solution and the schemes by eqs. (2.52) and (2.63) for $Pe = 5$, with one moving node ($S(x) = 0$). It is clear from the graph that the solution is improved. Numerical diffusion decreases.

2.7 Study of the scheme using the MNM

In this section we deal with the monotonicity and approximation of the scheme using the MNM. Based on the analytical form of the approximate solution to the problem between the nodes, which is obtained based on the MNM, it is possible to study the monotonicity and the type of approximation of the scheme.

2.7.1 Study of monotonicity

2.7.1.1 Scheme with central-difference approximation of the convective term

Consider eq. (2.40). Take a segment $[x_{i-1}, x_{i+1}] \subset [0,1]$ and any point $x \equiv x_i \in (x_{i-1}, x_{i+1})$. Let us consider the grid analogue of eq. (2.40):

$$\frac{u_{i+1} - u_{i-1}}{x_{i+1} - x_{i-1}} = \frac{2}{Pe(x_{i+1} - x_{i-1})}\left(\frac{u_{i+1} - u}{x_{i+1} - x} - \frac{u - u_{i-1}}{x - x_{i-1}}\right) + S(x). \tag{2.64}$$

If we set $x = (x_{i+1} + x_{i-1})/2$ we have a central-difference approximation. Here u_{i+1} is the approximate value of the solution at the point x_{i+1}, and u is the approximate value of the solution at the point x. To obtain a physically plausible solution in simple cases, we set $S(x) = 0$.

From eq. (2.64) we find

$$u = \frac{(x - x_{i-1})(2 - Pe(x_{i+1} - x))u_{i+1} + (x_{i+1} - x)(2 + Pe(x - x_{i-1}))u_{i-1}}{2(x_{i+1} - x_{i-1})}. \tag{2.65}$$

By changing the values x on the segment (x_{i-1}, x_{i+1}), we can determine the behavior of the solution. For given values x_{i+1}, x_{i-1}, u_{i-1}, u_{i+1}, eq. (2.65) is a parabola.

From eq. (2.65) we can obtain

$$\frac{u - u_{i-1}}{u_{i+1} - u_{i-1}} = \frac{(x - x_{i-1})(2 - Pe(x_{i+1} - x))}{2(x_{i+1} - x_{i-1})}. \tag{2.66}$$

A physically plausible solution will be obtained if $0 \le \frac{u - u_{i-1}}{u_{i+1} - u_{i-1}} \le 1$. This condition imposes the constraint $2 - Pe(x_{i+1} - x) \ge 0$. This condition is the monotonicity condition of the central-difference scheme for a nonuniform grid. In the case of a uniform grid, we have $2 \ge Pe \cdot h$. This condition is the well-known monotonicity condition [31]. For a coarse grid ($N = 2$, one movable node) with $Pe = 5$, the solution of the exact and approximate solutions is shown in Fig. 2.27.

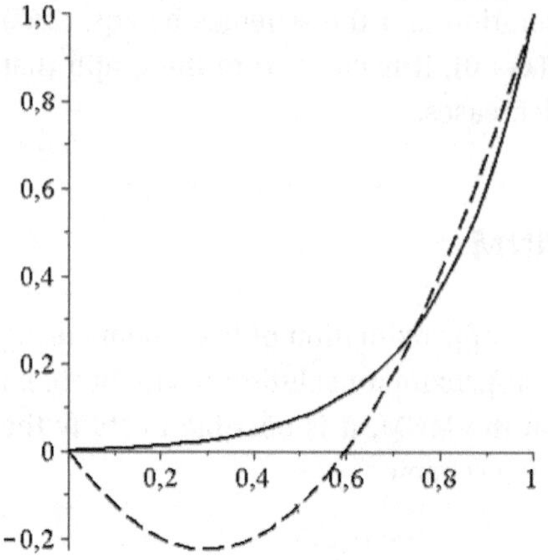

Fig. 2.27: Comparison of solutions in a coarse grid. The dotted curve is approximate, the solid curve is exact, and $Pe = 5$ ($\Phi_0 = 0$, $\Phi_1 = 1$).

In Fig. 2.27, the solid curve is the exact solution, and the dotted curve is the approximate solution obtained based on eq. (2.66). It is clear from the graph that scheme (2.66) does not give a physically plausible analytical solution. That is why scheme (2.66) at large Peclet numbers gives an oscillatory numerical solution. A plausible solution must have the same qualitative character as the exact solution. In the numerical solution, scheme (2.65) is implemented using a sweep, and for the stability of the sweep, the nodes are selected so that $2 - Pe(x_{i+1} - x_i) \geq 0$. For example, for a coarse grid (one nodal point), the plausibility condition gives $x_i \geq 0,6$. Indeed, at $Pe = 5$, it follows from (see Fig. 2.8) $2 - Pe(1 - x) \geq 0$.

For $Pe = 2$, the comparisons of the solution are shown in Fig. 2.28, which yields a physically plausible solution.

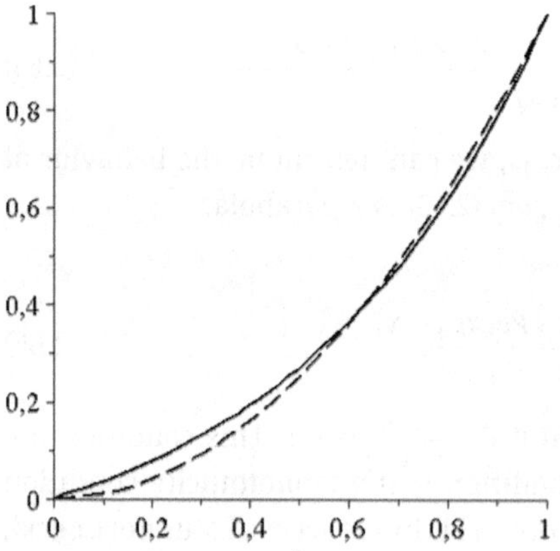

Fig. 2.28: Comparison of the solution in a coarse grid. The dotted curve is approximate, the solid curve is exact, and $Pe = 2$ ($\Phi_0 = 0$, $\Phi_1 = 1$).

2.7.1.2 Upstream scheme

Let us consider the difference analog of eq. (2.40), in which the convective term is approximated by a one-sided difference ratio (without a source):

$$\frac{u-u_{i-1}}{x-x_{i-1}}=\frac{2}{Pe(x_{i+1}-x_{i-1})}\left(\frac{u_{i+1}-u}{x_{i+1}-x}-\frac{u-u_{i-1}}{x-x_{i-1}}\right).$$

From here, we obtain

$$u=\frac{2(x-x_{i-1})u_{i+1}+(x_{i+1}-x)(2+Pe(x_{i+1}-x_{i-1}))u_{i-1}}{(x_{i+1}-x_{i-1})(2+Pe(x_{i+1}-x))}$$

or

$$\frac{u-u_{i-1}}{u_{i+1}-u_{i-1}}=\frac{2(x-x_{i-1})}{2(x_{i+1}-x_{i-1})(2+Pe(x_{i+1}-x))}. \tag{2.67}$$

For eq. (2.67), we always have $0\le\frac{u-u_{i-1}}{u_{i+1}-u_{i-1}}\le 1$. That is, the upstream scheme is always monotonic. Figure 2.28 shows a comparison of the exact and approximate analytical solutions ($Pe=5$). However, numerical diffusion occurs.

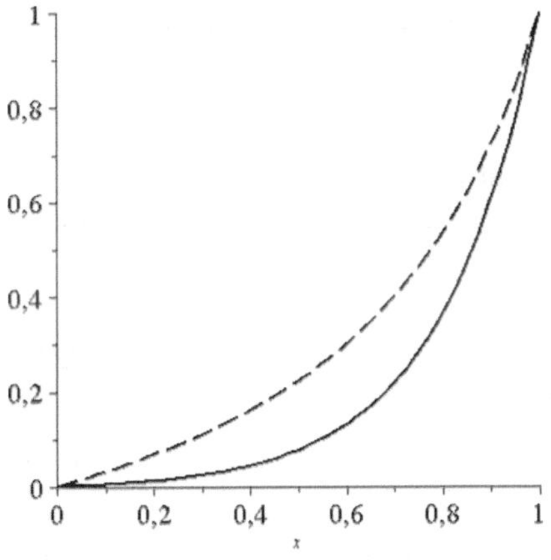

Fig. 2.29: Comparison of the solution in a coarse grid. The dotted curve is approximate, the solid curve is exact ($\Phi_0=0$, $\Phi_1=1$).

2.7.2 Calculating the residual

When replacing a DE with a difference equation, an approximation error occurs, which is usually expressed through the powers of the grid step [27]. In these cases, we speak of orders of approximation error. Using the MNM, it is possible to calculate the residual.

We introduce the approximation error as follows. Let the DE be given:

$$Lu=f, \tag{2.68}$$

where L is a differential operator, u is a known function, and f is an unknown function. Equation (2.28) is considered in a certain domain D with the corresponding boundary conditions. We replace eq. (2.68) with a difference equation:

$$L_h u_h = f_h, \tag{2.69}$$

where L_h is a difference operator, u_h is an unknown grid function, and f_h is an approximation of the function f at the grid nodes. Usually, the approximation error is given as follows [27]:

$$Q_h = L_h[u]_h - f_h, \tag{2.70}$$

where $[u]_h$ is the exact solution (2.28) at the grid nodes. Using the Taylor series, from eq. (2.70), we obtain that $Q_h = O(h^m)$, where h is the grid step and m is the approximation degree.

With the help of the MNM, we can introduce the approximation error as follows:

$$R_h = L_h\{u\}_h - f_h. \tag{2.71}$$

Here $\{u\}_h$ is a predefined continuous function by means of a moveable node. We demonstrate the approximate calculation of the approximation error of type (2.71) using the example of classical schemes of the convection-diffusion equation.

2.7.2.1 Scheme with central-difference approximation

Let us take a segment $[x_{i-1}, x_{i+1}] \subset [0,1]$ and any point $x \equiv x_i \in (x_{i-1}, x_{i+1})$. From eq. (2.25) we find

$$\frac{du}{dx} = \frac{2 - Pe(x_{i+1} + x_{i-1} - 2x)}{2} \cdot \frac{u_{i+1} - u_{i-1}}{x_{i+1} - x_{i-1}}, \tag{2.72}$$

$$\frac{d^2u}{dx^2} = Pe \cdot \frac{u_{i+1} - u_{i-1}}{x_{i+1} - x_{i-1}}. \tag{2.73}$$

Using formulas (2.72) and (2.73), the derivatives are reconstructed at any point of the segment $[0,1]$. From eqs. (2.72) and (2.73), we obtain

$$\frac{du}{dx} - \frac{1}{Pe}\frac{d^2u}{dx^2} = \Psi_1, \tag{2.74}$$

where

$$\Psi_1 = -Pe \cdot \frac{x_{i+1} + x_{i-1} - 2x}{2} \frac{u_{i+1} - u_{i-1}}{x_{i+1} - x_{i-1}}.$$

Equation (2.74) can be called a differential analog of difference equation (2.74), Ψ_1 there is a local approximation error of eq. (2.40) by difference equation (2.24) on the segment $[x_{i-1}, x_{i+1}]$. At the point $x = x_i$ we have $\Psi_1 = 0$, that is, difference equation (2.24) is a collocation scheme.

Using eq. (2.73), the approximation error can be written as

$$\Psi_1 = -\frac{x_{i+1} + x_{i-1} - 2x}{2}\frac{d^2u}{dx^2}.$$

Then eq. (2.40) takes the form:

$$\frac{du}{dx} - \left(Pe - \frac{x_{i+1} + x_{i-1} - 2x}{2}\right)\frac{d^2u}{dx^2} = 0. \tag{2.75}$$

Thus, difference equation (2.25) exactly approximates differential eq. (2.75) on the segment $[x_{i-1}, x_{i+1}]$ at $S(x) = 0$.

2.7.2.2 Upstream scheme

In the case of upstream scheme, we get

$$\frac{du}{dx} = \frac{2(2 + Pe(x_{i+1} - x_{i-1}))}{(2 + Pe(x_{i+1} - x))^2}\frac{u_{i+1} - u_{i-1}}{x_{i+1} - x_{i-1}}, \tag{2.76}$$

$$\frac{d^2u}{dx^2} = \frac{4Pe(2 + Pe(x_{i+1} - x_{i-1}))}{(2 + Pe(x_{i+1} - x))^3}\frac{u_{i+1} - u_{i-1}}{x_{i+1} - x_{i-1}}. \tag{2.77}$$

Calculate the approximation error

$$\Psi_2 = -\frac{2Pe(x_{i+1} - x)(2 + Pe(x_{i+1} - x_{i-1}))}{(2 + Pe(x_{i+1} - x))^3}\frac{u_{i+1} - u_{i-1}}{x_{i+1} - x_{i-1}}.$$

The differential analog of upstream scheme has the form

$$\frac{du}{dx} = \left(\frac{1}{Pe} + \frac{x_{i+1} - x}{2}\right)\frac{d^2u}{dx^2},$$

that is, for upstream scheme we have scheme diffusion with coefficient $(x_{i+1} - x)/2$.

2.8 On convergence of MNM

1. Consider the problem

$$\frac{d\Phi}{dx} = \frac{1}{Pe}\frac{d^2\Phi}{dx^2}, \qquad \Phi(0) = 0, \quad \Phi(1) = 1. \tag{2.78}$$

For this problem the difference scheme with $2^k - 1$ movable nodes has the form (1.29):

$$a_P^{(2^k-1)} U^{(2^k-1)} = a_E^{(2^k-1)} U_E^{(2^k-1)} + a_W^{(2^k-1)} U_W^{(2^k-1)}, \tag{2.79}$$

where

$$a_E^{(2^k-1)} = \frac{2^{2k+1}(1-\gamma_k)}{(1-x)(1-\gamma_k^{2k})}, \quad a_W^{(2^k-1)} = \frac{2^{2k+1}Pe(1-\tau_k)}{x(1-\tau_k^{2k})} + \frac{2^{2k+1}(1-\tau_k)}{x(1-\tau_k^{2k})},$$

$$a_P^{(2^\Pi-1)} = a_W^{(2^\Pi-1)} + a_E^{(2^\Pi-1)}. \; \tau_k = 2^k/(2^k+\sigma), \; \gamma_k = (2^k+\theta)/2^k, \; \sigma = Pe\,x,$$

$$\theta = Pe\,(1-x).$$

If we find from eq. (2.35) $U^{(2^k-1)}$ and pass to the limit at $k \to \infty$, we have

$$\lim_{k\to\infty} U^{(2^k-1)}(x) = \frac{e^{Pe\,x}-1}{e^{Pe}-1}.$$

The obtained limit coincides with the exact solution.

Chapter 3
Solution of the Cauchy problem

The idea of the movable node method is that discrete equations (in particular, a finite-difference equation) are transformed into a continuous form and on the basis of which an approximately analytical solution of the equation is obtained. Some possibilities for developing this approach are proposed here.

For the numerical solution of an ordinary differential equation with an initial condition, various varieties of the Euler method are widely used. Applying the movable node method to the Euler method, we obtain an approximately analytical expression for the solution of the Cauchy problem. This chapter presents the application of the movable node method to solving the Cauchy problem.

Let us consider the Cauchy problem

$$y' = \frac{dy}{dt} = f(t,y), \quad y(t_0) = f(t,y) \tag{3.1}$$

on segments $t \in [t_0, T]$.

3.1 Solution of the Cauchy problem

3.1.1 Explicit Euler method

For the numerical solution of problem (3.1), the explicit Euler method is often used (it is assumed $t_0 = 0$):

$$y_{k+1} = y_k + hf(t_k, y_k), \qquad k = 0, 1, \ldots, N-1. \tag{3.2}$$

Here $t_k = kh$, $h = \frac{T}{N}$.

To obtain an approximately analytical solution, we proceed as follows. Let t be an arbitrary point on the segments $[0,T]$. Considering t to be a movable node and setting $h = t - t_0$, following Euler's method, we have

$$U_1^1 = y_0 + tf(t_0, y_0). \tag{3.3}$$

Equation (31.3) is an approximate analytical solution to problem (3.1), that is, eq. (3.3) is a rough approximate solution (linear function). To refine the approximate solution (3.1), we divide the segment [0, t] into two parts: [0, t/2], [t/2,t].

Let us write an expression of type (1.3) for each segment:

$$U_1^2 = y_0 + \frac{t}{2} f(0, y_0),$$

https://doi.org/10.1515/9783111308173-004

$$U_2^2 = U_1^2 + \frac{t}{2} f\left(\frac{t}{2}, U_1^2\right).$$

The notation U_1^2 denotes an approximate solution on the segment [0, t/2] and U_2^2– to the segment – [t/2,t].

From here we can obtain

$$U_2^2 = y_0 + \frac{t}{2}\left[f(0,y_0) + f\left(\frac{t}{2}, y_0 + \frac{t}{2} f(0,y_0)\right)\right]. \tag{3.4}$$

Equation (3.4) is a more accurate approximation compared to eq. (3.3) for the entire segment [0, t].

Now we divide the segment [0, t] into n parts: [0, t/n], [t/n, $2t/n$], . . ., [$(n-1)t/n$,t]. For each segment we apply the Euler method, and as a result, we obtain

$$U_n^n = y_0 + \frac{t}{n}\left[f(0,y_0) + f\left(\frac{t}{n}, U_n^1\right) + \cdots + f\left(\frac{(n-1)t}{n}, U_n^{n-1}\right)\right]. \tag{3.5}$$

Here

$$U_n^i = y_0 + \frac{t}{n} f\left(\frac{it}{n}, U_n^{i-1}\right), \quad i = 1, 2, \ldots, n-1.$$

Calculating by formula (3.5), we obtain an approximate analytical solution to problem (3.1).

3.1.2 Implicit Euler method for a linear equation

Let the right-hand side of problem (3.1) be a linear function with respect to y:

$$y' = P(t)y + Q(t). \tag{3.6}$$

Here $P(t)$, $Q(t)$ are the given continuous functions.

We will proceed from the implicit Euler method. If we approximate (3.6) with one moving node, taking into account the initial condition, we have

$$U_1^1 = y_0 + tP(t)U_1^1 + tQ(t). \tag{3.7}$$

Hence

$$U_1^1 = \frac{y_0 + tQ(t)}{1 - tP(t)}. \tag{3.8}$$

Expression (3.8) is an approximate analytical solution to the Cauchy problem for eq. (3.6).

Let us now consider the segments [0, t/2], [t/2,t] and apply eq. (3.8) for each segment

$$U_1^2 = \frac{y_0 + \frac{t}{2}P\left(\frac{t}{2}\right)}{1 - \frac{t}{2}Q\left(\frac{t}{2}\right)}, \qquad U_2^2 = \frac{U_1^2 + tP(t)}{1 - tQ(t)}.$$

Hence, excluding U_1^2, we obtain

$$U_2^2 = \frac{y_0 + \frac{t}{2}Q(t)}{\left(1 - \frac{t}{2}P\left(\frac{t}{2}\right)\right)\left(1 - \frac{t}{2}P(t)\right)} + \frac{\frac{t}{2}Q(t)}{1 - \frac{t}{2}P(t)}. \tag{3.9}$$

If we divide the segment [0, t] into three parts: [0, t/3], [t/3,2t/3] and [2t/3,t] and write a formula of type (3.8) for each segment, we will eventually obtain

$$U_3^3 = \frac{1}{\left(1 - \frac{t}{3}P\left(\frac{t}{3}\right)\right)\left(1 - \frac{t}{3}P\left(\frac{2t}{3}\right)\right)\left(1 - \frac{t}{3}P(t)\right)}$$
$$\left[y_0 + \frac{t}{3}Q\left(\frac{t}{3}\right) + \frac{t}{3}Q\left(\frac{2t}{3}\right)\left(1 - \frac{t}{3}P\left(\frac{t}{3}\right)\right) + \frac{t}{3}Q(x)\left(1 - \frac{t}{3}P\left(\frac{t}{3}\right)\right)\left(1 - \frac{t}{3}P\left(\frac{2t}{3}\right)\right)\right]. \tag{3.10}$$

If we divide the segment with movable nodes into n parts, we obtain an analytical expression in the form:

$$U_n^n = \frac{1}{\prod\limits_{i=1}^{n}\left(1 - \frac{t}{n}P\left(\frac{it}{n}\right)\right)}\left[y_0 + \frac{t}{n}Q\left(\frac{t}{n}\right) + \frac{t}{n}\sum_{i=2}^{n} Q\left(\frac{it}{n}\right)\prod_{k=1}^{i-1}\left(1 - \frac{t}{n}P\left(\frac{kt}{n}\right)\right)\right]. \tag{3.11}$$

Formula (3.11) is an approximate analytical solution to the Cauchy problem for eq. (3.6).

3.1.3 Implicit Euler method for a nonlinear equation

Let us now consider the implicit Euler method for problem (3.1) using a movable node. To do this, we linearize the right-hand side of eq. (3.1) with respect to the variable y:

$$f(t,y) \approx f(t,y_0) + f'_y(t,y_0)(y - y_0). \tag{3.12}$$

Taking into account the linearization (3.12) and introducing the notations as follows:

$$P(t) = f'_y(t,y_0), \qquad Q(t) = f(t,y_0) - y_0 f'_y(t,y_0), \tag{3.13}$$

we obtain a linear equation of type (3.6). It should be noted that in the case of one moving node, we use formula (3.8). When using a multipoint moving node, formula (3.11) should be used.

Examples. Let us present some results of comparison with the exact solution.

1. Consider the Cauchy problem

$$y' = -ty + \sin t, \quad y(0) = 1. \tag{3.14}$$

Figures 3.1–3.3 show the results of solving problem (3.14) on the interval [0,5] using formula (3.5) for different partition numbers. The black lines correspond to the exact solution and the red ones to the approximate one.

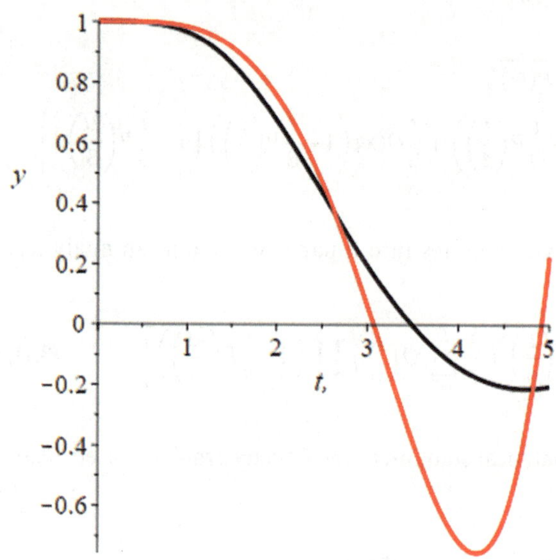

Fig. 3.1: Comparison of approximate solutions with the exact solution of problem (3.14) for $n = 3$.

Figure 3.1 shows that when the number of moved nodes is small ($n = 3$), then acceptable results are obtained at the beginning of the segment. However, at the end of the segment, the discrepancy between the exact and approximate solutions is significant. Figure 3.2 shows a comparison of the exact and approximate solutions with five moved nodes. In this case, we can state that good results are obtained on the segment [0, 4.5], and on the segment [4.5, 5], the discrepancy is significant.

Naturally, with a further increase in the number of moved nodes, the approximate solution will improve.

Figure 3.3 shows the results in the case of choosing with 10 movable nodes. Figure 3.3 shows a qualitative coincidence of the approximate solution with the exact one.

Figures 3.4–3.6 show the results of solving problem (3.14) on the interval [0,10] using formula (3.11) with different numbers of movable nodes.

Figure 3.4 shows that the implicit Euler method with a movable node gives qualitative agreement of the results (cf., Figs. 3.1 and 3.4). The results with five movable nodes are presented in Fig. 3.5.

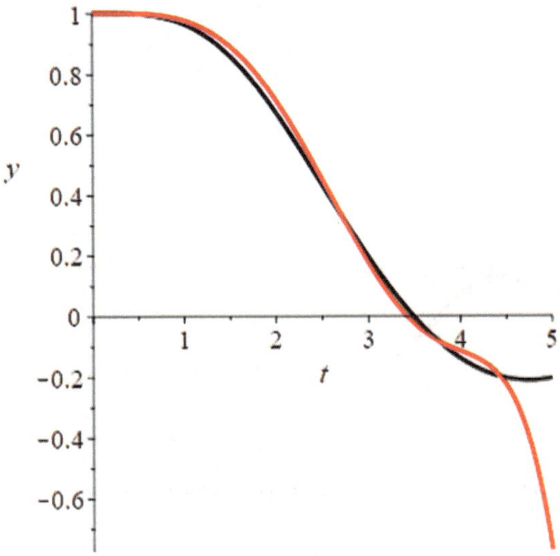

Fig. 3.2: Comparison of approximate solutions with the exact solution of problem (3.14) for $n = 5$.

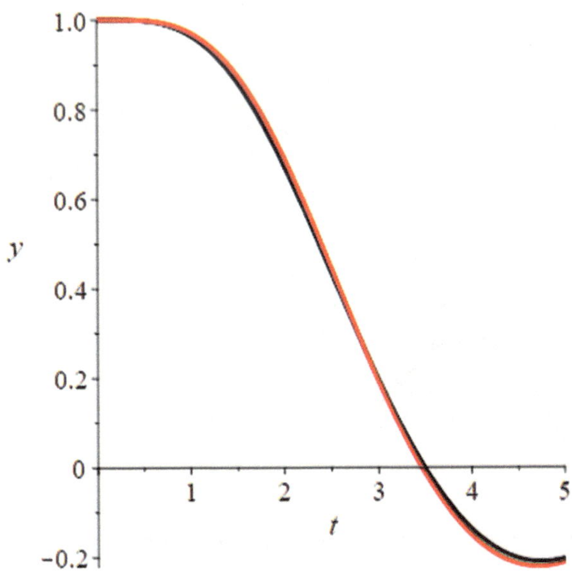

Fig. 3.3: Comparison of approximate solutions with the exact solution of problem (3.14) for $n = 10$.

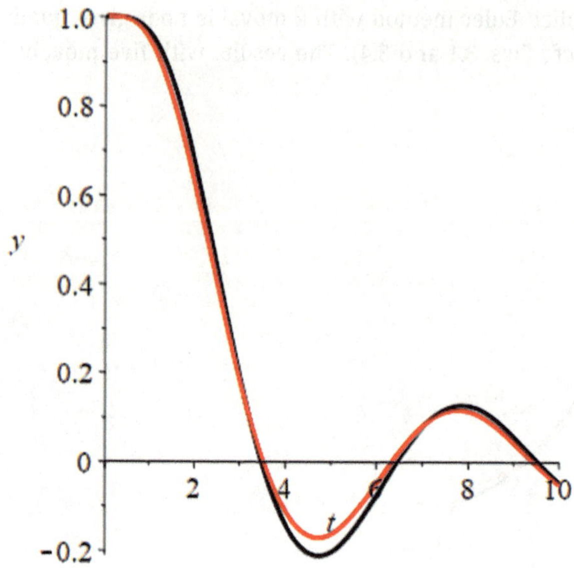

Fig. 3.4: Comparison of approximate solutions with the exact one of problem (3.14) for $n = 3$ on the interval [0, 10].

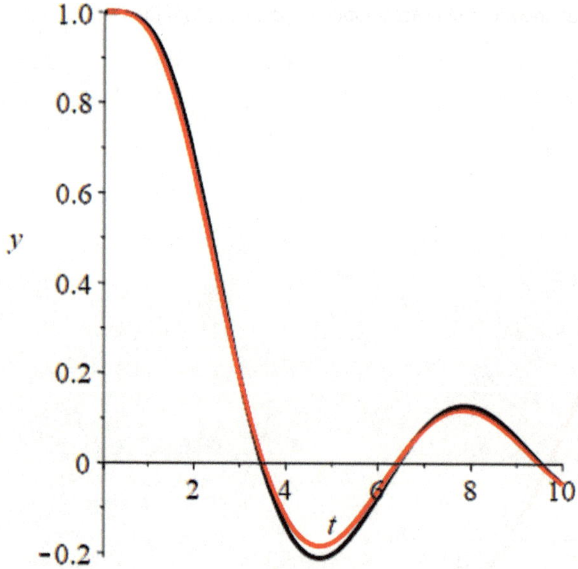

Fig. 3.5: Comparison of approximate solutions with the exact one of problem (3.14) for $n = 5$ on the interval [0, 10].

Comparison of Figs. 3.4 and 3.5 shows that with an increase in the number of movable nodes we approach the exact solution.

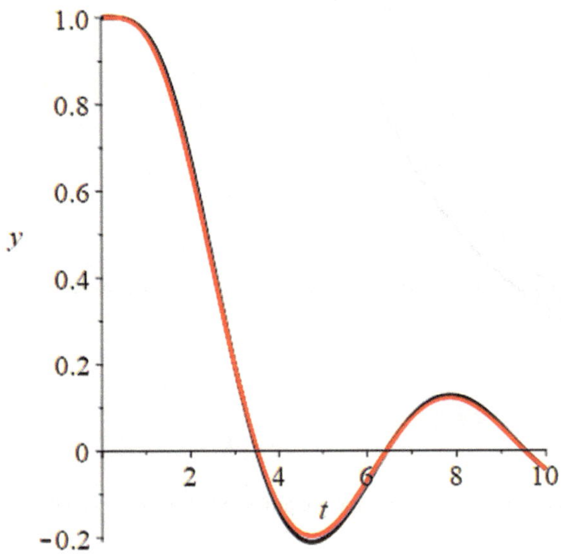

Fig. 3.6: Comparison of approximate solutions with the exact solution of problem (3.14) for $n = 10$ on the interval [0, 10].

Figure 3.6 shows an almost visual coincidence of the exact and approximate solutions with five movable nodes. Comparative analysis of Figs. 3.1–3.3 and Figs. 3.4–3.6, with the corresponding movable nodes, shows the advantage of the approximate solution obtained on the basis of the implicit Euler method.

3.1.4 Nonlinear problem

Let us consider the Cauchy problem:

$$y' = ty^2, \quad y(0) = 1. \tag{3.15}$$

Figure 3.7 shows a comparison of the solution of problem (3.15) using linearization (3.13) and according to formula (3.11) with five movable nodes.

The results of application with ten movable nodes are shown in Fig. 3.8.

Further increase in the number of movable nodes leads to convergence of the exact solution (Fig. 3.9).

Thus, the method of movable nodes can be successfully applied to the Cauchy problem. To clarify the solution, the number of movable nodes should be increased.

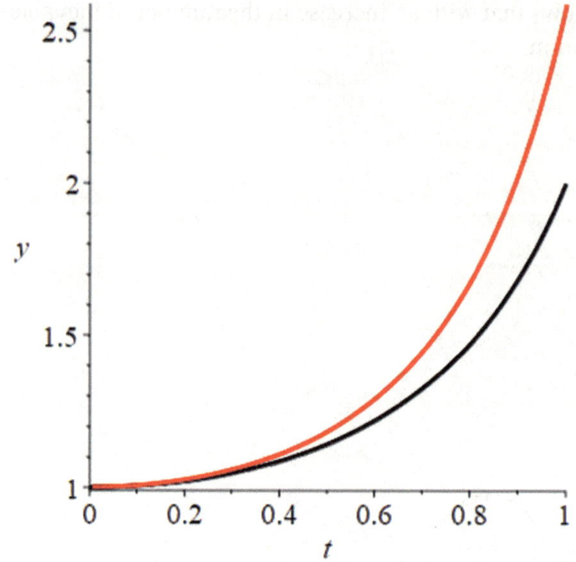

Fig. 3.7: Comparison of approximate solutions with the exact solution of problem (3.15) for $n = 5$ on the segment [0, 1].

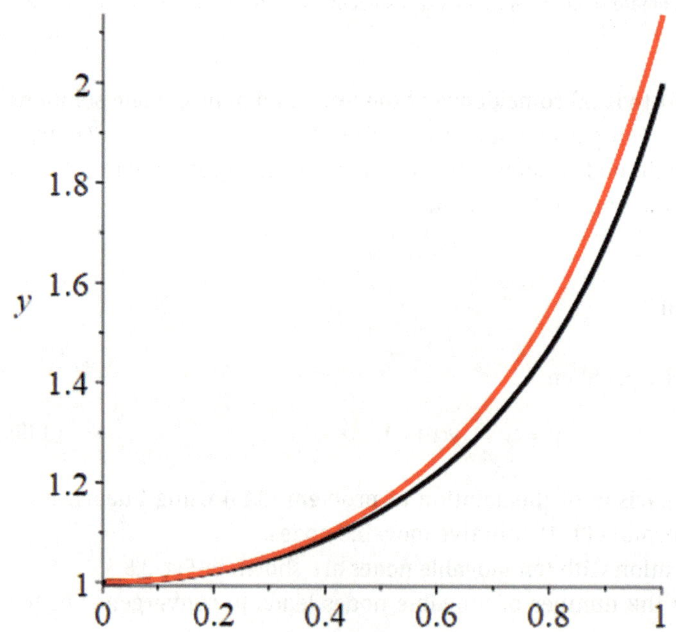

Fig. 3.8: Comparison of approximate solutions with the exact solution of problem (3.15) for $n = 10$ on the segment [0, 1].

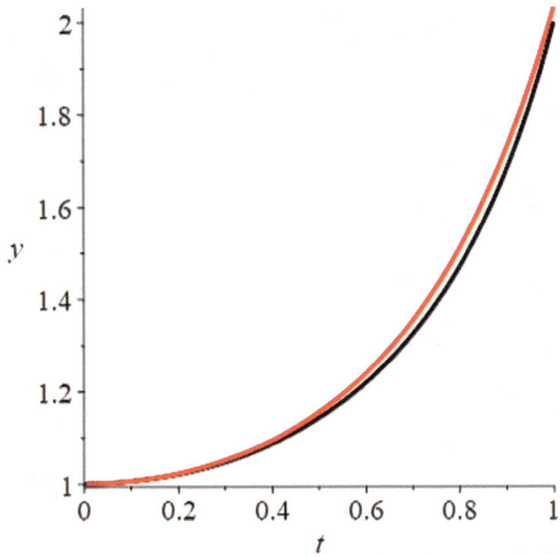

Fig. 3.9: Comparison of approximate solutions with the exact solution of problem (3.15) for $n = 15$ on the segment [0, 1].

3.2 On the convergence of the MNM

We will demonstrate the convergence of the MNM for the Cauchy problem on a model problem. Consider the Cauchy problem

$$\frac{du}{dx} = -u, \quad u(0) = 1. \tag{3.16}$$

Replace the derivative with the forward difference,

$$\frac{du}{dx} \approx \frac{U_1(x) - U_1(0)}{x - 0} = \frac{U_1(x) - 1}{x}. \tag{3.17}$$

In eq. (3.17), $U_1(x)$ is the approximate value of the unknown function at the moving point x, if the moving node is the only one.

Using eq. (3.17), we write the implicit difference equation:

$$\frac{U_1(x) - 1}{x} = -U_1(x). \tag{3.18}$$

Now let us take two moving nodes x and $x/2$. For these points, we write difference equations of the type (3.18):

$$\frac{U_2(x/2)-1}{x/2} = -U_2(x/2), \qquad \frac{U_2(x)-U_2(x/2)}{x-x/2} = -U_2(x). \tag{3.19}$$

Excluding in these equations $U_2(x/2)$ we obtain:

$$U_2(x) = \frac{1}{(1+x/2)^2}.$$

For three moving nodes $x/3$, $2x/3$ and x we obtain

$$U_3(x) = \frac{1}{(1+x/3)^3}.$$

If the number of nodes n, we obtain

$$U_n(x) = \frac{1}{(1+x/n)^n}. \tag{3.20}$$

If we tend the number of nodes to infinity, we obtain

$$\lim_{n\to\infty} U_n(x) = \lim_{n\to\infty} \frac{1}{(1+x/n)^n} = e^{-x}.$$

Thus, we obtain the exact solution of problem (3.16).

Chapter 4
Solution of initial-boundary value problems for a parabolic equation

The previous sections considered the application of the moving node method (MNM) for ordinary differential equations (ODEs). Here we consider the application of the MNM for parabolic equations.

An example of a problem leading to a parabolic partial differential equation is the problem of heat transfer along a long rod, described by the heat transfer (or diffusion) equation.

4.1 Approximate solution of a parabolic equation with one movable node

Let us consider a one-dimensional differential equation of parabolic type in the domain Ω ($0 < t < T$, $W < x < E$) [30]:

$$\frac{\partial u}{\partial t} = A\frac{\partial^2 u}{\partial x^2} + f(x,t), \tag{4.1}$$

with initial

$$u\,(x, t = 0,) = u^0(x) \tag{4.2}$$

and boundary conditions

$$u\,(x = W, t) = u_W(t) \;\; u(x = E, t) = u_E(t). \tag{4.3}$$

We assume that the solution of problems (4.1)–(4.3) exists and is unique.

For the numerical solution of problems (4.1)–(4.3) there are various difference schemes [23]. Let us consider various options for the difference approximation of a linear one-dimensional equation in space using a moving node.

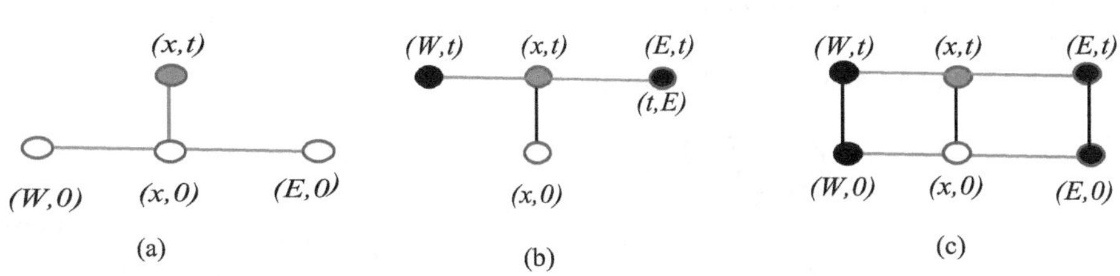

Fig. 4.1: Movable node template.

https://doi.org/10.1515/9783111308173-005

Figure 4.1 shows templates with one movable node. Figure 4.1(a) corresponds to the template according to the explicit scheme, Fig. 4.1(b) the implicit scheme and Fig. 4.1 (c) the Crank-Nicholson. In Fig. 4.1, the point $(x, t) \in \Omega$ refers to the movable node. The points $(x, 0)$, (W, t) and (E, t) are also movable nodes: the point $(x, 0)$ moves only along the x-axis, the point (W, t) moves along the left and the point (E, t) moves along the right boundary of the domain.

Using one movable node, we can obtain a rough analytical representation of the solution to problems (4.1)–(4.3).

Let us denote by $U(x,t)$. an approximate analytical solution to the problem obtained using the movable node method and using the boundary and initial conditions. Let $(x,t) \in \Omega$ be an arbitrary movable point. We approximate eq. (4.1) with an explicit scheme:

$$\frac{U(x,t) - U^0(x)}{t} = A\frac{2}{E-W}\left(\frac{U^0{}_E(E) - U^0(x)}{E-x} - \frac{U^0(x) - U^0(W)}{x-W}\right) + f(x,t). \tag{4.4}$$

In eq. (4.4) $U(x,t)$ there is an approximate analytical solution to the problem. When the point runs through Ω we obtain a solution in the region under consideration. From eq. (4.3) we obtain

$$U(x,t) = U^0(x) + A\frac{2t}{E-W}\left(\frac{U^0{}_E(E) - U^0(x)}{E-x} - \frac{U^0(x) - U^0(W)}{x-W}\right) + tf(x,t). \tag{4.5}$$

If we perform an approximation using an implicit scheme, we have

$$\frac{U(x,t) - U^0(x)}{t} = A\frac{2}{E-W}\left(\frac{U_E(t) - U(x,t)}{E-x} - \frac{U(x,t) - U_W(t)}{x-W}\right) + f(x,t). \tag{4.6}$$

Solving this, we obtain an approximate analytical solution to the problem in the case of an implicit scheme:

$$\begin{aligned} U(x,t) = {} & \frac{(E-x)(x-W)}{2At + (E-x)(x-W)}U^0(x) + \frac{2At[U_E(t)(x-W) + U_W(t)(E-x)]}{2At + (E-x)(x-W)} \\ & + \frac{(E-x)(x-W)t}{2At + (E-x)(x-W)}f(x,t). \end{aligned} \tag{4.7}$$

The Crank-Nicholson scheme has the form:

$$\begin{aligned} \frac{U(x,t) - U^0(x)}{t} = {} & \sigma A\frac{2}{E-W}\left(\frac{U_E(t) - U(x,t)}{E-x} - \frac{U(x,t) - U_W(t)}{x-W}\right) \\ & + (1-\sigma)A\frac{2}{E-W}\left(\frac{U^0(E) - U^0(x)}{E-x} - \frac{U^0(x) - U^0(W)}{x-W}\right) + f(x,t). \end{aligned} \tag{4.8}$$

From here we determine the approximate analytical solution using the Crank-Nicholson scheme:

$$U(x,t) = \frac{(E-x)(x-W)}{D}U^0(x) + B\frac{(x-W)U_E(t)-(E-x)U_W(t)}{D}$$
$$+C\frac{(x-W)(U^0(E)-U^0(x))-(E-x)(U^0(x)-U^0(W))}{D} + \frac{(E-x)(x-W)t}{D}f(x,t). \tag{4.9}$$

In expression (4.9), the following notation is introduced:

$$B = \frac{2\sigma At}{E-W};\quad C = \frac{2(1-\sigma)At}{E-W};\quad D = (E-x)(x-W) + B(E-W).$$

Let's consider test examples.

4.2 Examples of problem solutions

4.2.1 The first problem

Consider eq. (4.1) in the domain $\Omega = \{(x,t)\,|\,0 \le x \le 1,\ 0 \le t \le T\}$ with initial and boundary conditions:

$$U^0(x) = 2\sin\frac{\pi x}{h},\quad U_W(t) = 0,\quad U_E(t) = 0,$$

and with the right-hand side of the form: $f(x,t) = 0$. The exact solution of this problem is

$$U(x,t) = 2\sin\left(\frac{\pi x}{h}\right)\exp\left(-\frac{\pi^2 A}{h^2}t\right).$$

Figure 4.2 shows a comparison of the considered schemes for some sections. In these graphs, solid lines correspond to the exact solution. Figure 4.2(a) shows that the method of a moving node constructed according to the explicit one can be used only in the initial period; with distance from the initial point, a significant discrepancy occurs. Based on Fig. 4.2(b) and (c), the implicit schemes show a qualitative coincidence of the results. However, in the case of the moving node method, based on Fig. 4.2(d), which shows a comparison of the implicit scheme and Crank-Nicholson with the exact solution, it follows that the result is the opposite, the purely implicit scheme by the moving node gives a good result compared to the Crank-Nicholson scheme (the dotted line corresponds to the obtained purely implicit scheme, and the dotted-dotted line is obtained based on the

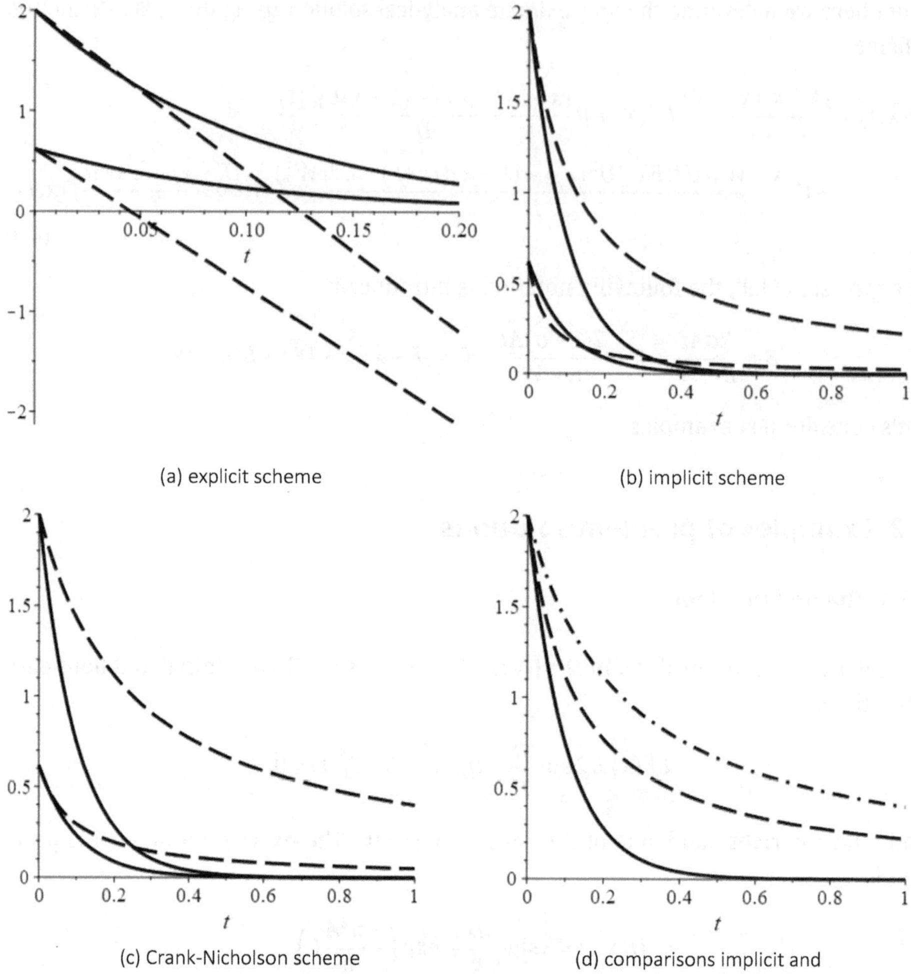

(a) explicit scheme

(b) implicit scheme

(c) Crank-Nicholson scheme

(d) comparisons implicit and according to the Crank-Nicholson scheme

Fig. 4.2: Comparisons of solutions for problem (4.1).

Crank-Nicholson scheme). It should also be mentioned that in the case of a nonuniform grid, the classical Crank-Nicholson scheme is a first-order scheme.

Figure 4.3 shows a comparison of the mentioned scheme of the difference between the exact and approximate solutions, the difference between the exact and approximate solution is plotted vertically. Figure 4.2 shows the advantage of the purely implicit scheme of the moving node.

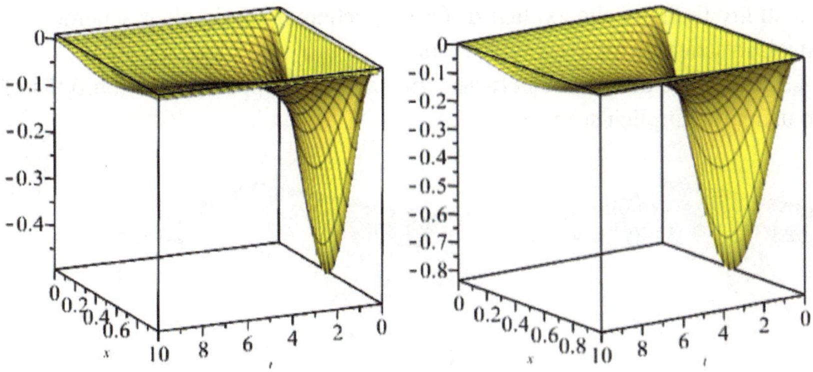

(a) purely implicit scheme (b) Crank-Nicolson scheme

Fig. 4.3: Difference between exact and approximate solutions of problem (4.2).

4.2.2 Second problem

Consider eq. (4.1) in the domain $\Omega = \{(x,t) \mid 0 \le x \le 1, 0 \le t \le 10\}$ with the conditions:

$$U^0(x) = x, \; U_W(t) = 0, \; U_E(t) = e^{-t}, \; f(x,t) = -x\,e^{-t}. \tag{4.10}$$

Exact solution of the problem $U(x,t) = x\,e^{-t}$. Figure 4.3 shows a comparison of the exact and approximate solutions over the cross section $x = 0.5$ and $x = 0.2$. Solid lines correspond to the exact solution.

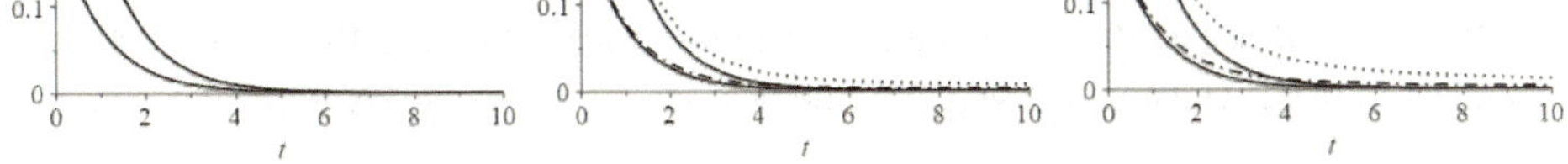

(a) explicit scheme (b) implicit scheme (c) Crank-Nicolson scheme

Fig. 4.4: Solution of eq. (4.1) with parameters (4.19).

Figure 4.4 shows a significant discrepancy between the exact and approximate solutions far from the initial point. Apparently, the beginning of the discrepancy is observed outside the stability limits of the explicit difference scheme. The implicit scheme and the Crank-Nicolson scheme give acceptable results.

Figure 4.5 shows the difference between the exact and approximate solutions of the problem using the implicit schemes.

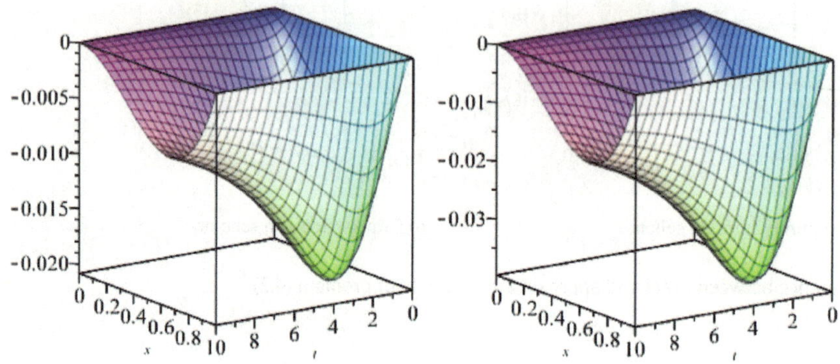

(a) implicit scheme

(b) Crank-Nicolson scheme (σ=0.5)

Fig. 4.5: The difference between the exact and approximate solutions of eq. (4.1) with parameters (4.19).

Note that the difference between the exact and approximate solutions gives a result that is close in form. In this example, a similar result follows as in the case of example 1: a purely implicit scheme for a moving node gives a good result compared to the Crank-Nicolson scheme (cf. Figs. 4.4 and 4.5).

To clarify this issue, calculations were carried out with a change in the Crank-Nicolson scheme: the value of the source term in formula (4.9) $f(x,t)$ was replaced by $f(x, t/2)$. Figure 4.6 shows a comparison of this change (for greater clarity, the graphs corresponding to the section $x = 0.5$ are shown in the figures). In Fig. 4.6, the solid line corresponds to the exact solution, the dotted line to the purely implicit scheme, the dashed line to the Crank-Nicholson scheme with the source term $f(x,t/2)$ and the dash-dotted line to the Crank-Nicholson scheme with the source term $f(x,t)$.

In the initial period ($0 < t < 4$), the closest to the exact solution is the Crank-Nicholson scheme for $f(x,t/2)$. Outside this segment, the best scheme is the purely implicit scheme.

4.2.3 The third problem

Consider eq. (4.1) in the domain $\Omega = \{(x,t)\ |\ 0 \le x \le 1,\ 0 \le t \le 10\}$ with initial and boundary conditions:

$$U^0(x) = \sin \pi x + x^2,\ U_W(t) = 0,\ U_E(t) = 1,$$

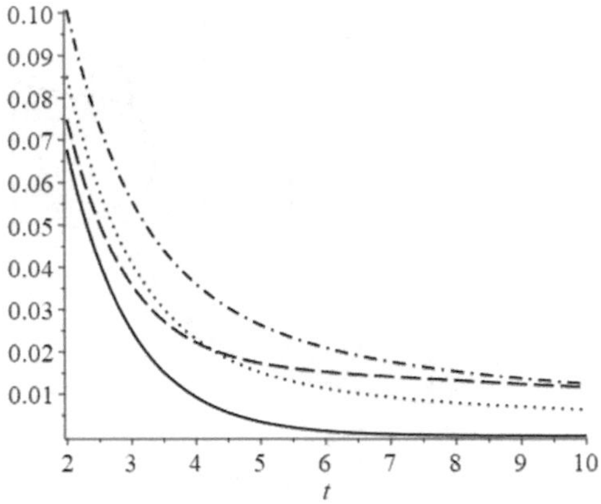

Fig. 4.6: Comparison of implicit schemes.

and with the right-hand side of the form:

$$f(x,t) = (\pi^2 - 1)\sin \pi x \cdot e^{-t} - 2.$$

Exact solution of this problem is

$$u(x,t) = \sin \pi x\, e^{-t} + x^2.$$

For this problem, the same tendencies are observed with respect to the comparison of schemes; that is, the explicit scheme does not give a picture of the solution close to the exact one, the closest result is for the implicit scheme. Figure 4.7 shows a comparison of the exact and approximate solutions.

Comparison of Fig. 4.7(a) and (b) visually shows the closeness of the exact and applied solutions. From comparison of Fig. 4.7(c) and (d) it follows that the implicit scheme is much more accurate than the Crank-Nicolson scheme.

4.2.4 Unsteady flow of viscous fluid between parallel walls

As a practical example, consider the unsteady flow of viscous fluid between parallel walls. Let the viscous fluid fills the entire space between horizontal planes located at a certain distance from each other. Let the lower plane be motionless all the time, and the upper one begins to move to the right at $t = 0$ a constant speed at a moment. We neglect the effect of gravity and assume that the pressure is constant everywhere. We assume that the flow is directed parallel to the ox-axis. Then the equation of motion of a viscous fluid in dimensionless variables has the form [4]:

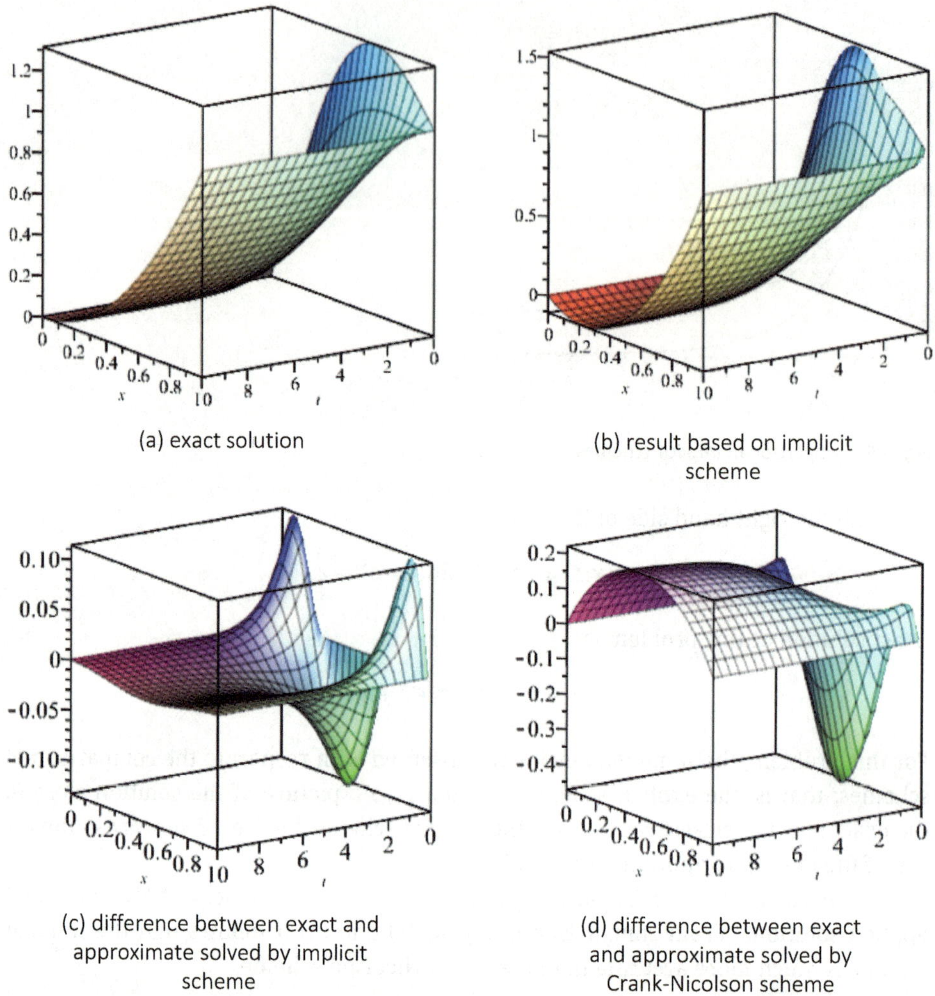

(a) exact solution

(b) result based on implicit scheme

(c) difference between exact and approximate solved by implicit scheme

(d) difference between exact and approximate solved by Crank-Nicolson scheme

Fig. 4.7: Comparison of solutions.

$$\frac{\partial u}{\partial t} = \frac{\partial^2 u}{\partial y^2}. \tag{4.11}$$

The exact solution of the equation under the conditions:

$$u(0,y) = 0, \quad u(t,0) = 0, \quad u(t,1) = 1$$

has the form:

$$u = y + \frac{2}{\pi}\sum_{k=1}^{\infty}\frac{(-1)^k}{k}\sin(k\pi y)\exp(-k^2\pi^2 t). \tag{4.12}$$

It is easy to see that the analytical solution obtained on the basis of the explicit scheme does not give a physically plausible solution.

To obtain an implicit scheme of the moving node, we replace in eq. (4.11) with the difference relation:

$$\frac{u_1(t,y)-u_1(0,y)}{t-0} = \frac{2}{1-0}\left[\frac{u_1(t,1)-u_1(t,y)}{1-y} - \frac{u_1(t,y)-u_1(t,0)}{y-0}\right].$$

From here, taking into account the boundary conditions, we obtain an approximate solution in the form:

$$u_1(t,y) = \frac{2ty}{y(1-y)+2t}. \tag{4.13}$$

In the case of a Crank-Nicholson-type scheme, we obtain the following solution:

$$u_2(t,y) = \frac{2t\sigma y}{y(1-y)+2\sigma t}. \tag{4.14}$$

Note that from the exact solution (4.12) we can obtain a stationary solution $\lim_{t\to\infty} u(t,y) = y$, showing the linear distribution of velocity over the cross section of the pipe.

A similar result is obtained using approximate analytical solutions using formulas (4.13) and (4.14); that is, $\lim_{t\to\infty} u_1(t,y) = y$, $\lim_{t\to\infty} u_2(t,y) = y$.

Figure 4.8 shows the exact and approximate solution obtained using the implicit scheme, which shows the closeness of the solutions.

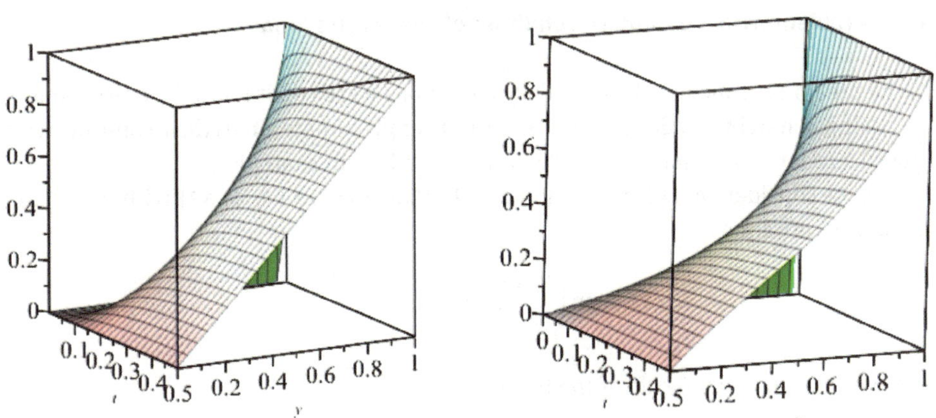

Fig. 4.8: Exact and approximate solution of the problem.

Figure 4.9 more clearly shows the nature of the change in speed over time for specific values of the variable y.

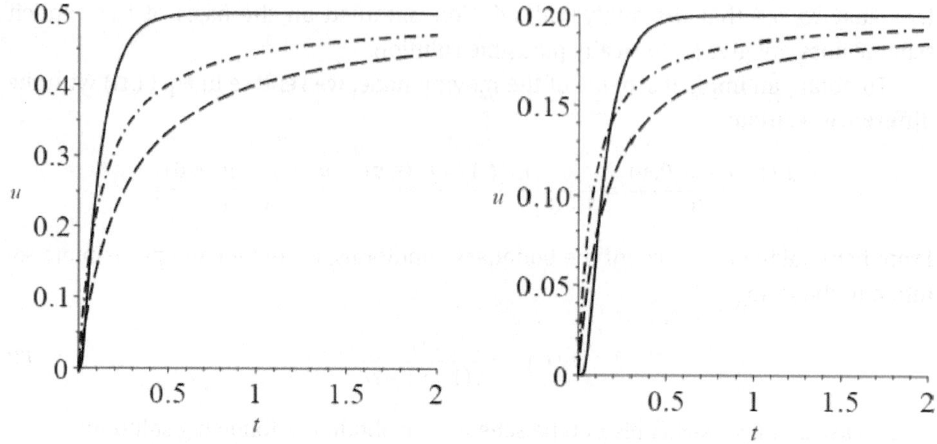

(a) results for y = 0.5 (b) results for y = 0.2

Fig. 4.9: Comparison of the results of different schemes.

The solid lines in Fig. 4.9 correspond to the exact solution, the dotted lines refer to the Crank-Nicholson scheme and the dotted-dotted line refers to the implicit scheme. In this example, the assertion takes place that the results for the implicit scheme give a more refined solution.

4.2.5 Problem with a boundary condition of the second kind

In the previous problems, the boundary conditions were conditions of the first kind.

Using a moving node, we can obtain an approximate analytical solution with other boundary conditions. Let us demonstrate this with examples.

Let us consider eq. (4.1) in the domain $\Omega = \{(x,t) \mid 0 \le x \le 1, 0 \le t \le 1\}$ with initial and boundary conditions:

$$U^0(x) = \sin\left(\frac{3\pi x}{2}\right), \quad U_W(t) = 0, \quad \left.\frac{\partial u}{\partial x}\right|_{x=1} = t$$

and with the right-hand side of the form:

$$f(x,t) = x.$$

The exact solution to this problem is of the form:

$$u(x,t) = x \cdot t + \exp\left(-\left(\frac{3\pi}{2}\right)^2\right) \sin\left(\frac{3\pi x}{2}\right).$$

To obtain an approximate solution, we use the implicit scheme (4.6). To use formula (4.7), which expresses an approximate solution, we must first determine the right boundary condition. First, we approximate the right boundary condition on the segments (1/2, 1):

$$\frac{U_E(t) - U(1/2,t)}{1/2} = t.$$

Hence,

$$U_E(t) = \frac{t}{2} + U(1/2,t). \tag{4.15}$$

To determine the value $U(1/2,t)$, we use formula (4.7) assuming $x = 1/2$, then

$$U(1/2,t) = \frac{1}{4l+1} U^0(1/2) + \frac{4t[U_E(t) \cdot 1/2]}{8l+1} + \frac{t}{8t+1}\frac{1}{2}.$$

Putting the value of the unknown in the right boundary obtained in eq. (4.15) in the form:

$$U_E(t) = \frac{8t+1}{8t+2} t + \frac{1}{4t+1} \sin\left(\frac{3\pi}{4}\right) + \frac{t}{8t+2}. \tag{4.16}$$

We have found an approximate form of the right boundary, and now it is easy to obtain a solution based on eq. (4.7):

$$U(x,t) = \frac{(1-x)x}{2t+(1-x)x} \sin\left(\frac{3\pi x}{2}\right) + \frac{2t[U_E(t)x]}{2t+(1-x)x} + \frac{(1-x)xt}{2t+(1-x)x} x. \tag{4.17}$$

Figure 4.9 shows the graphs of the exact and approximate solutions of the problem under consideration.

From Fig. 4.10, it is clear that the approximate solution is a good representation of the exact solution.

4.3 Approximate solution for the initial boundary value problem of a parabolic equation by the method of lines

This section proposes a method for obtaining an approximate solution for a linear parabolic equation with initial boundary conditions. In [9], an approximate analytical solution was obtained for a linear parabolic equation using the method of moving

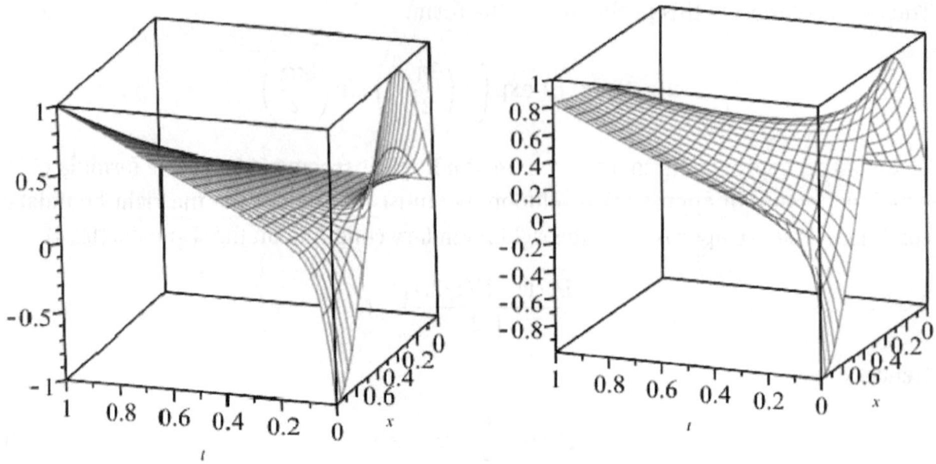

(a) exact solution (b) approximate solution

Fig. 4.10: Exact and approximate solution of the problem.

nodes. Here we will consider some schemes obtained based on the method of moving nodes in the context of the method of lines.

Consider the initial boundary value problems (4.1)–(4.3):

Using an implicit scheme, we obtained an approximate solution in the form of eq. (4.7).

In the method of lines for solving eqs. (4.1)–(4.3), only the second derivative is approximated at each nodal point by a difference ratio, and as a result, we obtain a system of ordinary equations, with the number of equations equal to the number of internal grid nodes with respect to the variable x. The resulting system of equations is solved by numerical methods for ODEs.

We apply the method of moving nodes to problems (4.1)–(4.3). Approximating the second derivative in eq. (4.1) with a moving node, we pass to an ODE. If we approximate the derivative with respect to the variable t in eq. (4.1), we again obtain a second-order ODE.

4.3.1 Approximation with respect to the variable *t*

In eq. (4.1), we approximate the derivative with respect to the variable t. Then we obtain ($U(x,t)$ approximate solution):

$$\frac{U(x,t)-U(x,0)}{t} = A\frac{d^2U}{dx^2} + f(x,t). \tag{4.18}$$

Hence,

$$At\frac{d^2U}{dx^2} - U(x,t) = -t\cdot f(x,t) - u(x,0). \tag{4.19}$$

In numerical analysis [11–13], the method of lines is widely used. Here, eq. (4.19) is considered as an ODE with respect to the variable x, and the variable t is used as a parameter.

We solve the nonhomogeneous differential eq. (4.19). If A does not depend on x, the exact solution of the homogeneous eq. (4.19) has the form

$$U^* = C_1\exp(-\sqrt{p}x) + C_2\exp(\sqrt{p}x),$$

where $p = \frac{1}{At}$ is set, and $U^*(x,t)$ – the solution of the homogeneous equation corresponds to eq. (4.19). We denote by $\bar{U} = Q(x,t)$ a particular solution of eq. (4.19). Taking into account the boundary conditions, we obtain

$$U(x,t). = C_1\exp(\sqrt{p}x) + C_2\exp(-\sqrt{p}x) + Q(x,t). \tag{4.20}$$

The constant coefficients C_1 and C_2 are determined based on the boundary condition (4.2) ($W = 0$):

$$C_1 = \frac{\exp(-\sqrt{p}b)\left(U^0(0) - Q(0,t)\right) - \left(U^0(E) - Q(E,t)\right)}{\exp(-\sqrt{p}E) - \exp(\sqrt{p}E)},$$

$$C_2 = \frac{\left(U^0(E) - Q(E,t)\right) - \exp(\sqrt{p}E)\left(U^0(0) - Q(0,t)\right)}{\exp(-\sqrt{p}E) - \exp(\sqrt{p}E)}.$$

The solution obtained by formula (4.20) is an approximate analytical solution of problems (4.1)–(4.3) with partial approximation by variable t based on the moving node.

4.3.2 Approximations by variable *x*

Now we approximate eq. (4.1) only by variable x:

$$\frac{dU(x,t)}{dt} = \sigma\frac{b}{2}\left[\frac{U(b,t) - U(x,t)}{b-x} - \frac{U(x,t) - U(0,t)}{x}\right] + f(x,t). \tag{4.21}$$

Hence

$$\frac{dU(x,t)}{dt} + p(x)\cdot U = q(x,t). \tag{4.22}$$

Here it is set

$$p(x) = \frac{2A}{(E-x)x},\quad q(x,t) = \frac{2A}{E}\cdot\frac{u_E(t)\,x + u_0(t)(E-x)}{x(E-x)}. + f(x,t).$$

We have obtained a linear differential equation of the first order. We solve eq. (4.22). Considering x as a parameter and t as an independent variable, we obtain the solution:

$$U(x,t) = \exp\left(-\int p(x)dt\right)\left(\int_0^t q(x,t)\exp\left(\int p(x)dt\right)dt + U^0(x)\right). \tag{4.23}$$

If A does not depend on t

$$\exp\left(-\int p(x)dt\right) = \exp\left(-\frac{2At}{x(E-x)}\right).$$

In this case, the solution has the form:

$$U(x,t) = \exp\left(-\frac{2At}{x(E-x)}\right)\left[U^0(x) + \int_0^t q(x,t)\exp\left(\frac{2At}{x(E-x)}\right)dt\right]. \tag{4.24}$$

If A and the boundary conditions do not depend on t, we obtain a solution in simple form:

$$U(x,t) = U^0(x)\exp(-p\cdot t) + E(xu_E + (E-x)u_0)(1-\exp(-p\cdot t)). \tag{4.25}$$

Examples. Let us consider some results of comparison with the exact solution.

1. Let us consider the problem of solving eq. (4.1) with the conditions:

$$1).\ \varphi(x) = \sin(x) + \frac{x}{\pi},\ E = \pi,\ u_0(t) = 0,\ u_E(t) = 1. \tag{3.26}$$

Figure 4.11 shows the solution to problems (4.1) and (4.3) with three approaches to approximating derivatives.

Comparisons of the results of approximate solutions with exact ones were carried out using the mean square error according to the formula:

$$S = \frac{\sum_{i,j=1}^{MN}(u_{ij}-U_{ij})^2}{\sqrt{NM}}.$$

Here it is assumed $u_{ij} = u(t_i, x_j),\ U_{ij} = U(t_i, x_j),\ N = 100,\ M = 100.$

The calculated values of the standard deviation with the help of three approximation methods are, respectively, equal to (a) $S = 0.067477$; (b) $S = 0.051968$; and (c) $S = 0.048134$.

2. The solution of eq. (4.1) is considered under the following conditions:

$$2).\ \varphi(x) = \sin(x) + \sin(2x),\ b = \pi,\ u_0(t) = 0,\ u_b(t) = 1. \tag{4.27}$$

Figure 4.12 shows the exact and approximate solutions with three approximation methods.

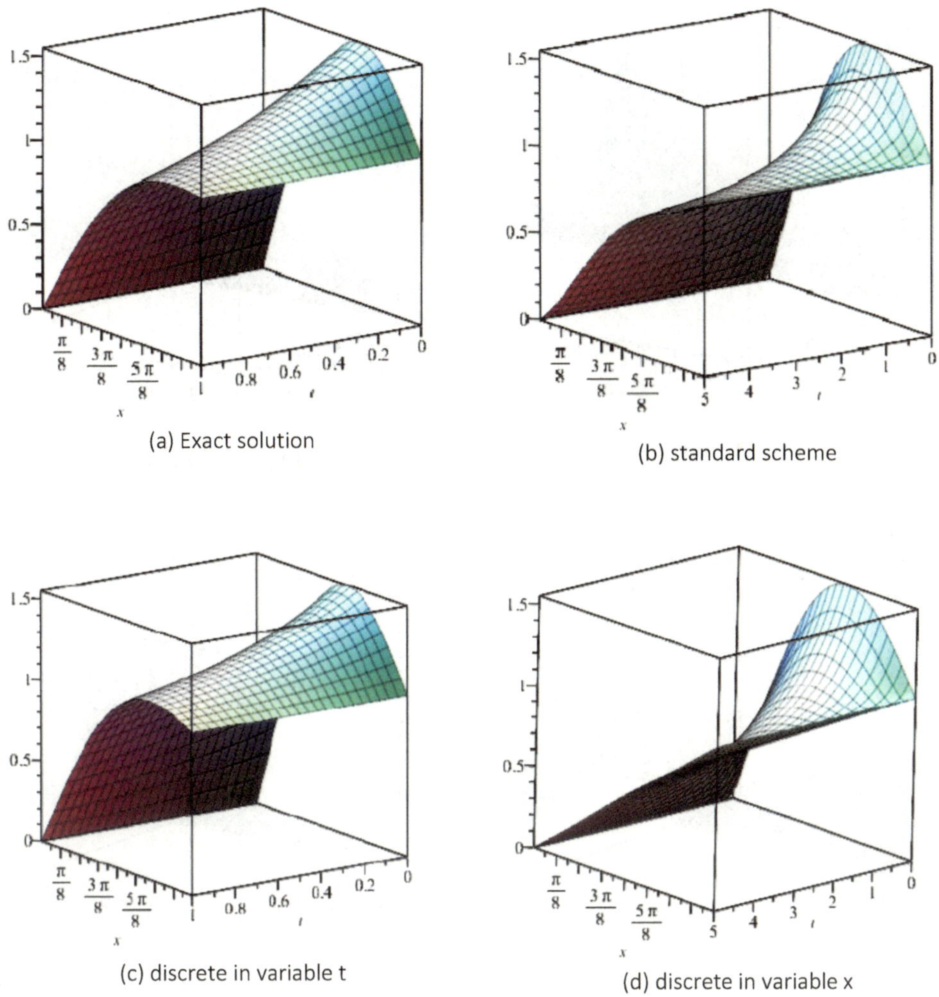

Fig. 4.11: Comparisons of solutions to problems (4.1) and (4.3).

Standard deviation: (a) $S = 0.1915033$; (b) $S = 0.070304$; (c) $S = 0.139150$.

3. Consider the problem of unsteady flow of a viscous fluid between parallel walls using the method of lines.

Replace (4.11) with the following equation:

$$\frac{du_2}{dt} = \frac{2}{1-0}\left[\frac{u_2(t,1) - u_2(t,y)}{1-y} - \frac{u_2(t,y) - u_2(t,0)}{y-0}\right]. \tag{4.28}$$

Considering y as a parameter in eq. (4.28) and solving it according to the rules for the exact solution of a first-order ODE, we obtain

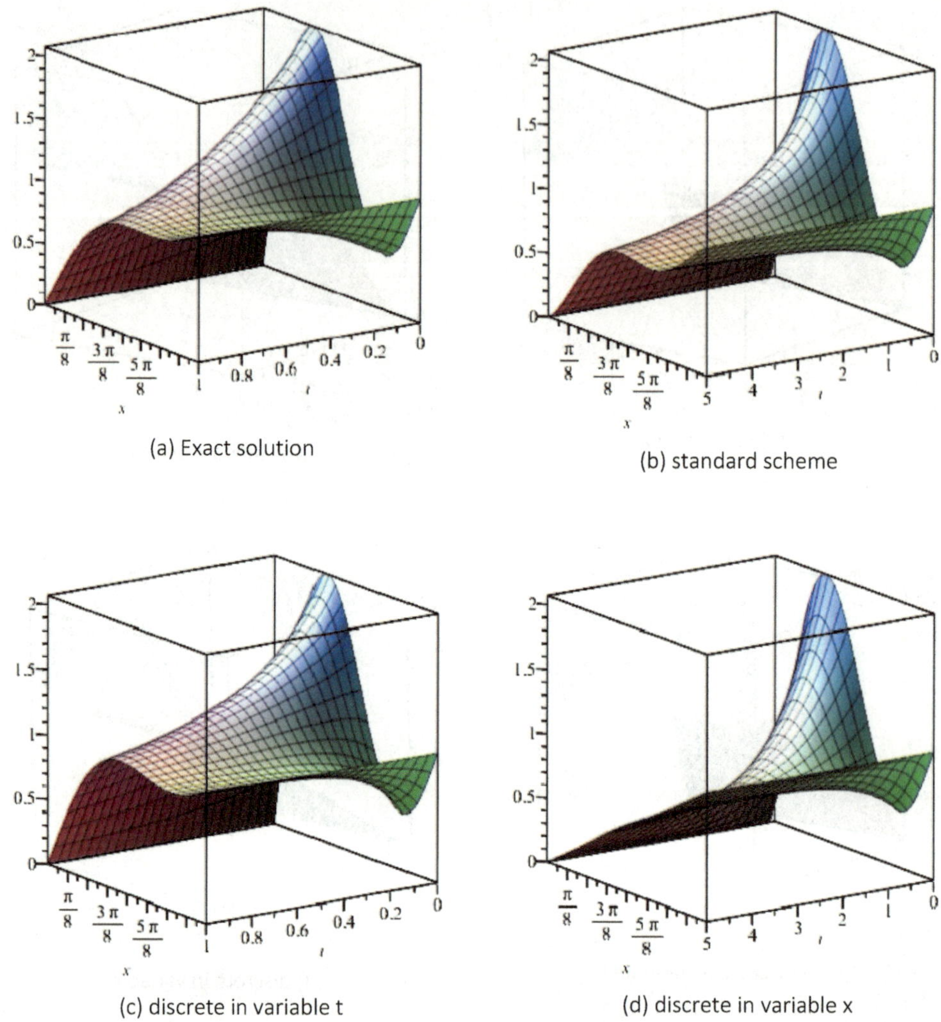

Fig. 4.12: Comparison of solutions to problem (4.27).

$$u_2(t,y) = y\left(1 - \exp\left(\frac{2t}{y(1-y)}\right)\right). \tag{4.29}$$

Now we make an approximation in time, that is:

$$\frac{d^2u_2}{dx^2} = \frac{u_3(t,x) - u_3(0,x)}{t}.$$

Solving the resulting second-order ODE, taking into account the boundary and initial conditions, we obtain

$$u_3(t,y) = \frac{\exp\left(-\frac{x}{\sqrt{t}}\right) - \exp\left(\frac{x}{\sqrt{t}}\right)}{\exp\left(-\frac{1}{\sqrt{t}}\right) - \exp\left(\frac{1}{\sqrt{t}}\right)}. \tag{4.30}$$

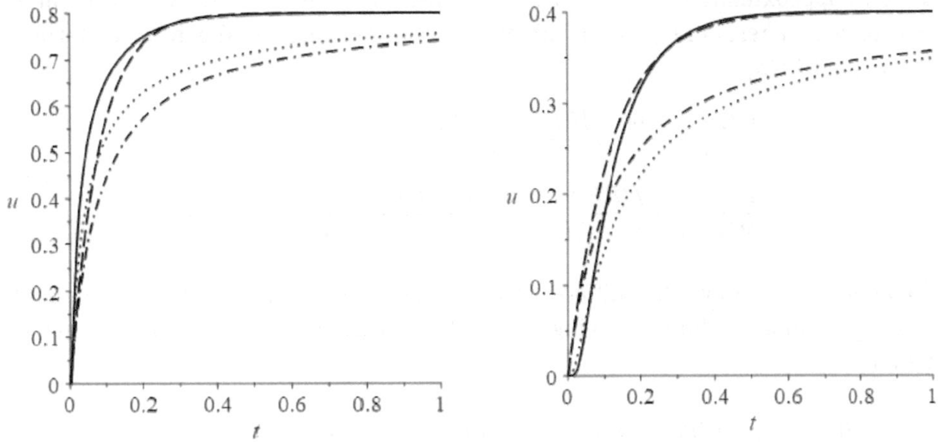

(a) velocity profile at y = 0.8 (b) velocity profile at y = 0.4

Fig. 4.13: Velocity profiles obtained using the moving node method.

Figure 4.13 shows the development of velocity in two positions. In the graphs, solid curves refer to the exact solution. The dotted graphs correspond to formula (4.12), the dotted and dashed lines correspond to eq. (4.30) and the dashed lines correspond to eq. (4.29). This shows that partial approximation by x gives a good result. Apparently, this is due to the fact that the process occurs more quickly in time, and it is necessary to approximate this member in the equation that changes slowly.

4.4 Nonstationary convection-diffusion equation

If the processes contain both diffusion and convective terms, then such processes are described by nonstationary differential convection-diffusion equations. Such problems are often encountered in various fields of computational mathematics, physics and chemistry.

Let us consider an equation with constant coefficients (Pe is the Peclet number characterizing the convection intensity):

$$\frac{\partial u}{\partial t} + \frac{\partial u}{\partial x} = \frac{1}{Pe}\frac{\partial^2 u}{\partial x^2} + f(x,t), \tag{4.31}$$

under conditions (4.2) and (4.3).

4.4.1 Approximation of the convection-diffusion equation

There are various schemes for approximating convection-diffusion equations, especially those associated with approximation by a convective term. Similar methods arise in approximation associated with the moving node method. We approximate the convective term of eq. (4.31) with an upstream scheme and the diffusion term in the standard way:

$$\begin{aligned}&\frac{U(x,t)-U(x,0)}{t}+\frac{U(x,t)-U(W,t)}{x-W}\\&=\frac{1}{Pe}\frac{2}{(E-W)}\left(\frac{U(E,t)-U(x,t)}{E-x}-\frac{U(x,t)-U(W,t)}{x-W}\right)+f(x,t).\end{aligned}\tag{4.32}$$

Equation (4.32) is an implicit difference scheme with a moving node. In eq. (4.32), $U(x,t)$ is an approximate solution. Solving eq. (4.32) for $U(x,t)$, we obtain an approximate solution:

$$\begin{aligned}U(x,t)=\frac{1}{M}\Big\{&u^0Pe(E-x)(x-W)+2t[u_E(x-W)+u_W(E-x)]/(E-W)\\&+Pe(E-x)u_W+t\cdot Pe(E-x)(x-W)f(x,t)\Big\}.\end{aligned}\tag{4.33}$$

Here we introduce the notation:

$$M=Pe(E-x)(x-W)+2t+Pe(E-x)],$$

and use the initial (4.2) and boundary conditions (4.3):

$$u^0=u^0(x),\ u_E=u(x=E,t),\ u_W=u(x=W,t).$$

In the case where we use the central-difference ratio for the convective term,

$$\begin{aligned}&\frac{U(x,t)-U(x,0)}{t}+\frac{U(E,t)-U(W,t)}{E-W}\\&=\frac{1}{Pe}\frac{2}{(E-W)}\left(\frac{U(E,t)-U(x,t)}{E-x}-\frac{U(x,t)-U(W,t)}{x-W}\right)+f(x,t),\end{aligned}$$

we obtain a solution in the form:

$$\begin{aligned}U(x,t)=\frac{1}{M_1}\Big\{&u^0Pe(E-x)(x-W)+2t[u_E(x-W)+u_W(E-x)]/(E-W)\\&+\frac{Pe(E-x)(x-W)(u_E-u_W)}{E-W}+t\cdot Pe(E-x)(x-W)f(x,t)\Big\}.\end{aligned}\tag{4.34}$$

In this case

$$M_1 = Pe(E-x)(x-W)+2t.$$

There is also an exponential type scheme [25], in which, using the profile of the exact solution of the stationary equation without a source term, a joint approximation of the convective and diffusion terms of the convection and diffusion equations is obtained. Using this approach for eq. (4.31), we can construct the following type of scheme with a movable node:

$$\frac{U(x,t)-U(x,0)}{t}+\frac{2(U(E,t)-U(W,t))}{E-W}$$
$$=\frac{2}{(E-W)}\left(\frac{U(E,t)-U(x,t)}{\exp(Pe(E-x))-1}-\frac{U(x,t)-U(W,t)}{\exp(Pe(x-W))-1}\right)+f(x,t).$$

From here we can obtain an approximate solution in the form:

$$U(x,t)=\frac{(E-W)A\cdot B}{(E-W)A\cdot B+2tA\cdot B+2t(A+B)}\left\{u^0+\frac{2tU_a}{E-W}+\frac{2t[u_E B+u_W A]}{(E-W)A\cdot B}+t\cdot f\right\}. \quad (4.35)$$

For the sake of simplicity, the following notations are introduced here:

$$A=\exp(Pe(E-x))-1, \quad B=\exp(Pe(x-W))-1.$$

4.4.2 Test examples

***Example* 1.**
Let us consider eq. (4.31) in the domain $\Omega=\{(x,t)\ |0\le x\le 1,\ 0\le t\le T\}$ with initial and boundary conditions:

$$u^0(x)=x^2+x, \ u_W(t)=0, \quad u_E(t)=1+\exp(-Pe\cdot t),$$

and with the right-hand side of the form:

$$f(x,t)=(1+Pe\cdot x)\exp(-Pe\cdot t)+2x-2/Pe.$$

The exact solution of this problem has the form:

$$U(x,t)=x^2+\exp(-Pe\cdot t).$$

A comparison of the approximate solution obtained using eq. (4.33) with the exact one is shown in Fig. 4.14 for $Pe = 0.1$ obtained using the upstream scheme, and Fig. 4.15 shows the results for eq. (4.34) (solid curves are the exact solution, dotted curves are the approximate solution, and the graphs correspond to the sections $x = 0.1; x = 0.5; x = 0.8$).

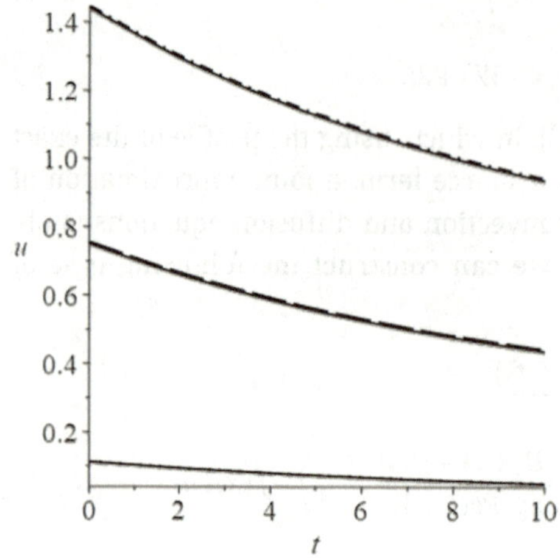

Fig. 4.14: Comparison of solutions (by formula (4.33) $Pe = 0.1$).

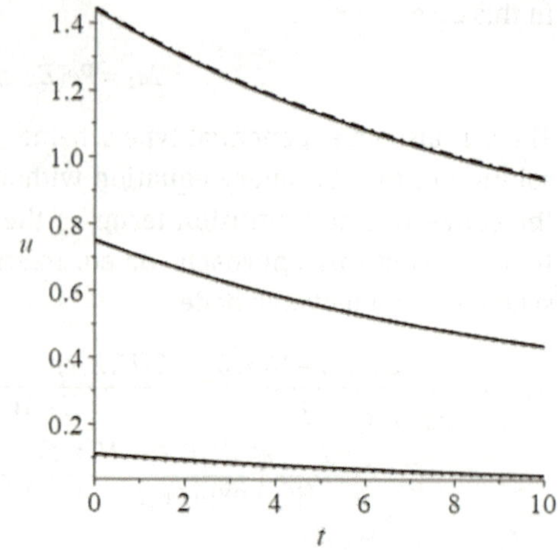

Fig. 4.15: Comparison of solutions (by formula (4.34) $Pe = 0.1$).

Figures 4.14 and 4.15 show good agreement of the results for both schemes. For a more detailed presentation, Figs. 4.16 and 4.17 show the differences in the accuracy of the exact and approximate solutions, showing that the solution obtained by eq. (4.34) is more acceptable for these parameters.

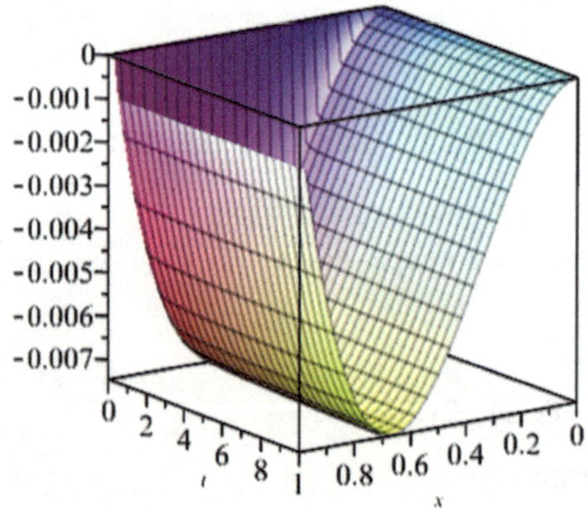

Fig. 4.16: Difference between the exact and approximate solutions (by formula (4.33) $Pe = 0.1$).

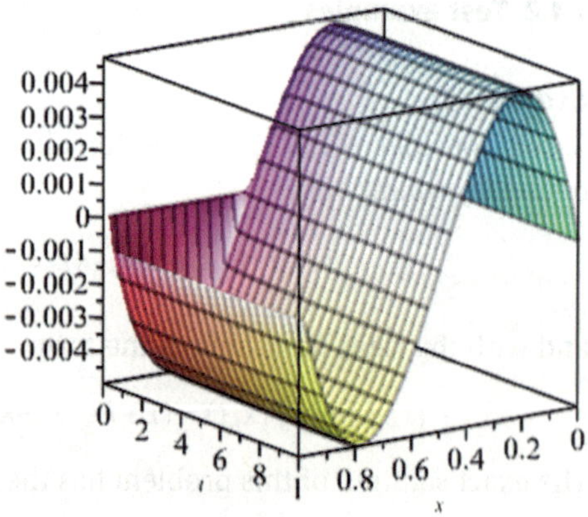

Fig. 4.17: Difference between the exact and approximate solutions (by formula (4.34) $Pe = 0.1$).

Increasing the Peclet number significantly affects the accuracy of the approximate solutions (Fig. 4.18).

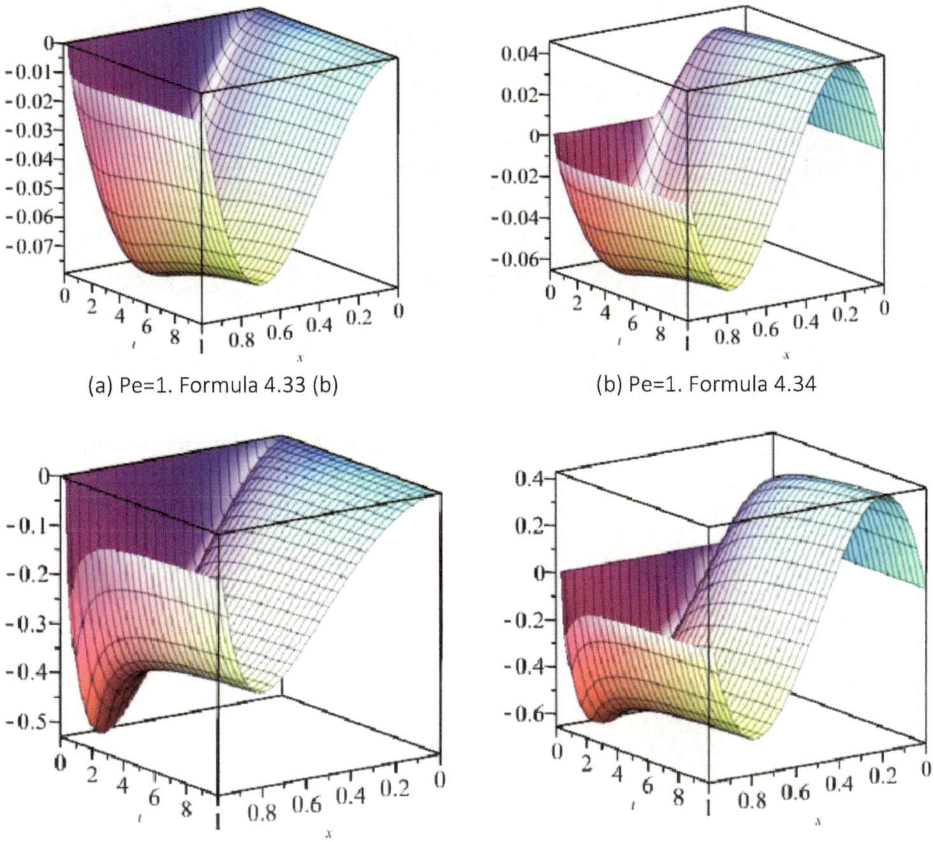

(a) Pe=1. Formula 4.33 (b)

(b) Pe=1. Formula 4.34

(c) Pe=10. Formula 4.33

(d) Pe=10. Formula 4.34

Fig. 4.18: Solution accuracy depending on the Peclet number.

***Example* 2.** Consider eq. (4.31) in the domain $\Omega = \{(x,t) \mid 0 \le x \le 1,\ 0 \le t \le T\}$ with initial and boundary conditions:

$$u^0(x) = \sin(x), \quad u_W(t) = \exp(-t)\sin(-t), \quad u_E(t) = \exp(\pi - t),$$

and with the right-hand side of the form:

$$f(x,t) = \left(\frac{1}{Pe} - 1\right) \exp(-t)\sin(x - t).$$

The exact solution to this problem is of the form:

$$u(x,t) = \exp(-t)\sin(x - t).$$

For small values of the Peclet number, the results from the formulas give close results. Figure 4.19 shows the exact and approximate solutions.

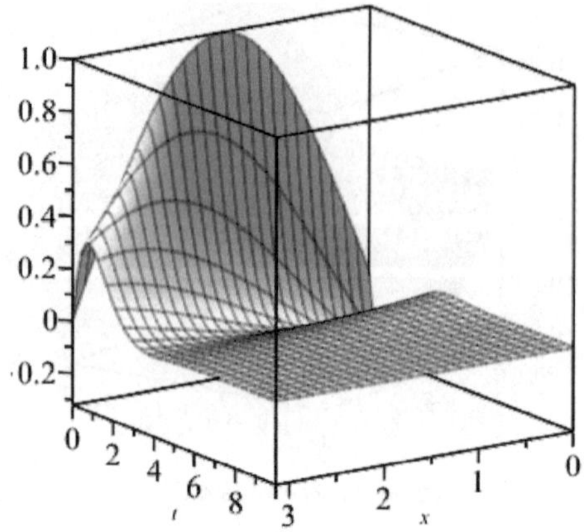

(a) Exact solution

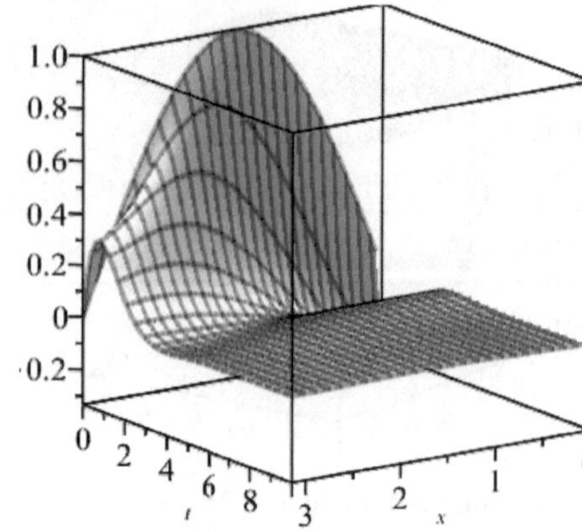

(b) Approximate solution (by (4.33), Pe=0.1)

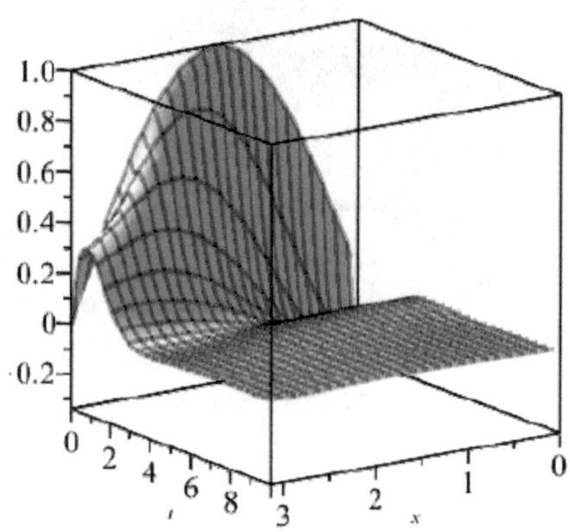

(c) Approximate solution (by (4.34), Pe=0.1)

Fig. 4.19: Exact and approximate solution of Example 2.

With increasing Peclet number, an increase in the discrepancy with the exact solution is observed. In this case, the results obtained by formula (4.34) begin to deteriorate compared to formula (4.33).

***Example* 3. Comparison with the exponential scheme.** A comparative analysis of the solution (4.35) was performed. For $Pe = 0.1$, the difference in the solution is similar to that shown in Fig. 4.18, that is, it coincides with the result obtained by the central scheme. Comparisons of Figs. 4.18(a), (b) and 4.20(a) show an improvement in the solution when using the exponential scheme. The same tendency is preserved at $Pe = 10$

(comparison of Figs. 4.17(c), 4.18(d), and 4.20(b)). A further increase in the Peclet number will lead to the unsuitability of the solution obtained by formula (4.34).

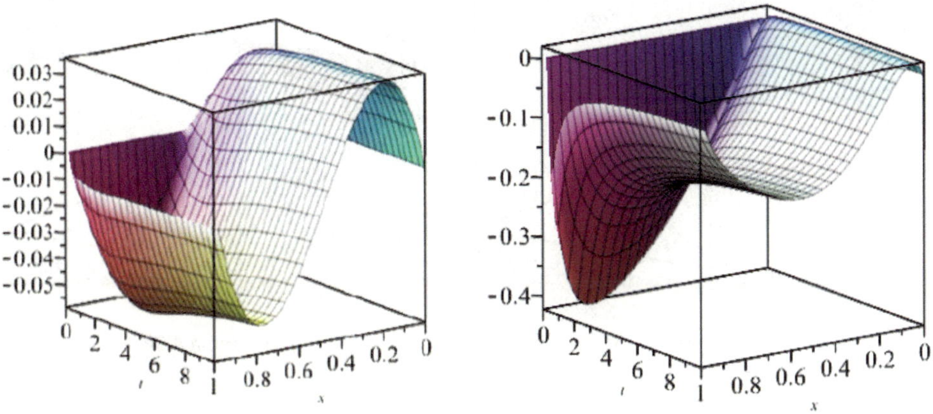

(a) Pe=1. Formula 4.35

(b) Pe=10. Formula 4.35

Fig. 4.20: Difference between exact and approximate solutions.

4.5 Improving the solution of a parabolic equation with multipoint movable nodes

4.5.1 Approximate solution of a parabolic equation based on a multipoint movable node

The use of the movable node method for a parabolic equation in various aspects in a single-point version is described in the previous paragraphs. The possibility of improving the solution using partial approximation of differential operators based on the line method approach is shown. The purpose of this section is to develop a multipoint version of the movable node method for problems (4.1)–(4.3).

In the case of two-point boundary value problems [9, 31], movable nodes were used: one-point, $\omega_1 = \{0,\ x,\ b\}$, three-point – $\omega_3 = \{0,\ x/2,\ x,\ (b+x)/2,\ b\}$, etc. In the given templates, nodes 0 and b coincide with the boundary point of the problem, and x is a movable node. Index ω denotes the number of movable nodes.

In Section 4.1 in Ω, one movable node is used, $\omega_{11} = \omega_{x1} \times \omega_{t1}$, where $\omega_{x1} = \{0, x, b\}$, $\omega_{t1} = \{0,\ t,\ T\}$. Now index ω denotes the number of movable nodes by variable, for example, and ω_{x3} indicates that three movable nodes are used by variable x. We introduce additional movable nodes:

$$\omega_{x3} = \{0,\ x/2,\ x,\ (b+x)/2,\ b\};\quad \omega_{t2} = \{0,\ t/2,\ t\};\quad \omega_{t3} = \{0,\ t/3,\ 2t/3,\ t\}.$$

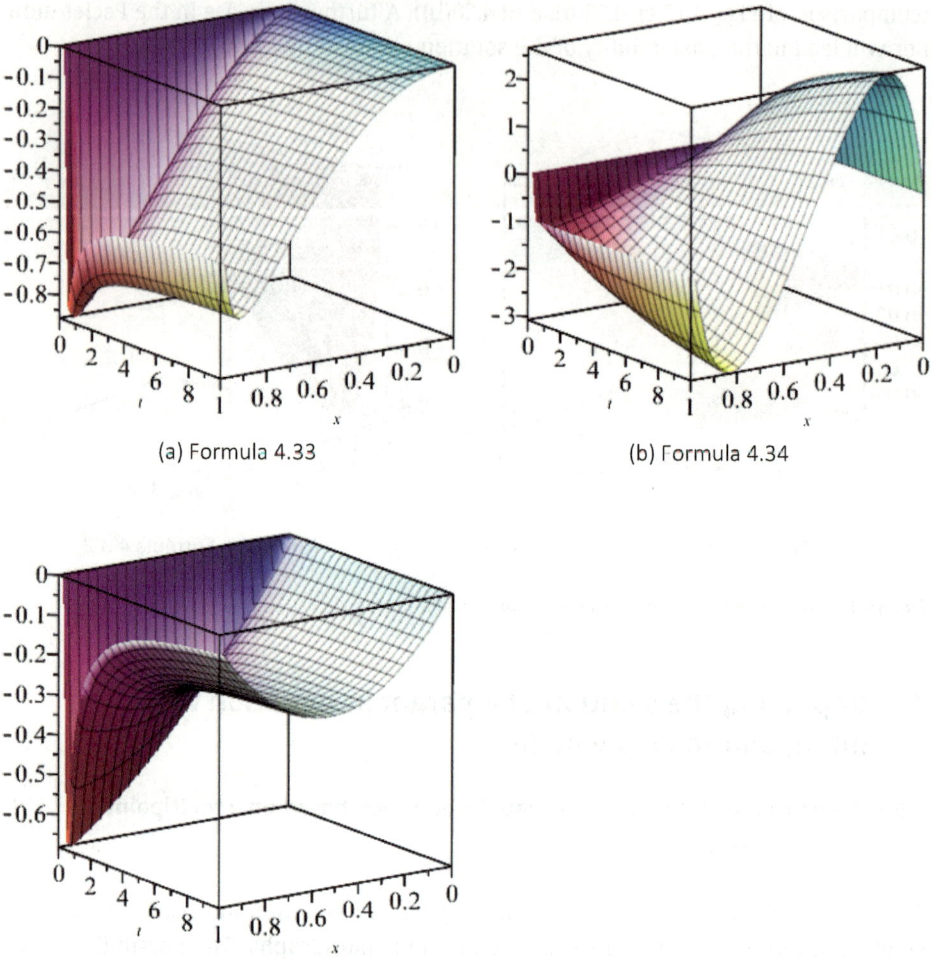

(a) Formula 4.33

(b) Formula 4.34

(c) Formula 4.35

Fig. 4.21: Difference between exact and approximate solutions (*Pe* = 100).

Let us first consider the approximation of eq. (4.1) with one movable node, that is, in ω_{11} with an implicit scheme:

$$\frac{U(x,t)-U(x,0)}{t}=\sigma\frac{b}{2}\left[\frac{U(b,t)-U(x,t)}{b-x}-\frac{U(x,t)-U(0,t)}{x}\right]+f(x,t). \tag{4.36}$$

In the equation $U(x,t)$ is the approximate value of the unknown function at the point (x,t). Using the boundary and initial conditions, we have

$$U(x,0)=u(x,0)=\varphi(x),\quad U(0,t)=u(0,t)=u_0(t),\quad U(b,t)=u(b,t)=u_b(t).$$

Solving eq. (4.36) by taking into account the boundary and initial conditions, we obtain

$$U(x,t) = \frac{x(b-x)}{2\sigma t + x(b-x)}\varphi(x) + \frac{2\sigma t[u_b(t)x + u_0(t)(b-x)]}{b(2\sigma t + x(b-x))} + \frac{x(b-x)tf(x,t)}{x(b-x)+2\sigma t}. \tag{4.37}$$

Equation (4.37) is an approximate analytical solution to problems (4.1)–(4.3).

The approximate solution satisfies conditions (4.2) and (4.3).

To improve the solution, we add movable nodes. In the pattern approximation $\omega_{31} = \omega_{x3} \times \omega_{t1}$, we obtain three types of movable nodes: $(x/2,t)$, (x,t) and $((b+x)/2,t)$. For each movable node, we use three implicit schemes of the type (4.36). From this system of equations, we determine the solution, which has the form:

$$U(x,t) = \left\{\frac{M\varphi(x)}{2\sigma t + M} + \frac{4\sigma t M}{b(2\sigma t + M)}\left[\frac{(b-x)\varphi((b-x)/2)}{2M_2} + \frac{(x-a)}{2M_1}\right]\right.$$
$$\left. + \frac{2(2\sigma t)^2 M}{b(2\sigma t + M)}\left[\frac{u_b(t)}{(b-x)(2\sigma t + M_2)} + \frac{u_0(t)}{x(2\sigma t + M_1)}\right]\right\}: \tag{4.38}$$
$$\left\{1 - \frac{2(2\sigma t)^2 M}{(b-a)(2\sigma t + M)}\left[\frac{1}{(b-x)M_2} + \frac{1}{xM_1}\right]\right\}.$$

In eq. (4.38), the notations are introduced:

$$M = x(b-x)/4, \quad M_1 = 2\sigma t + (x-a)^2/2, \quad M_2 = 2\sigma t + (b-x)^2/2.$$

Formula (4.38) is derived for $f(x,t) = 0$.

Now, let us consider improving the solution by adding alternating nodes with respect to the variable t. For brevity, we introduce the notations:

$$P(t) = 2\sigma t + x(b-x), \quad S_1(t) = \frac{x(b-x)}{P(t)}, \quad S_2(t) = \frac{2\sigma t[u_b(t)x + u_0(t)(b-x)]}{P(t)},$$
$$S_3 = S_1(t)\, t \cdot f(x,t).$$

Based on these notations, eq. (4.37) can be written as follows:

$$U(x,t) = S_1(t)\varphi(x) + S_2(t) + S_3(t). \tag{4.39}$$

Note that the variable x is not explicitly specified in the expressions $P(t)$, $S_1(t)$, $S_2(t)$, $S_3(t)$; there is no need for this.

Let us consider the pattern $\varpi_{12} = \varpi_{x1} \times \varpi_{t2}$, that is, movable nodes; $(x,t/2)$ and (x,t). Let us write formula (4.39) for these nodes:

For node $(x,t/2)$:

$$U(x,t/2) = S_1(t/2)\varphi(x) + S_2(t/2) + S_3(t/2) \tag{4.40}$$

and for node (x,t),

$$U(x,t) = S_1(t)U(x,t/2) + S_2(t) + S_3(t). \tag{4.41}$$

Substituting expression (4.40) into (4.41), we obtain

$$U(x,t) = S_1(t)[S_1(t/2)\varphi(x) + S_2(t/2) + S_3(t/2)] + S_2(t) + S_3(t). \tag{4.42}$$

If we use three-point moving nodes by variable t, we have

$$\begin{aligned} U(x,t) = S_1(t)\{S_1(2t/3)[S_1(t/3)\varphi(x) + S_2(t/3) + S_3(t/3)] + S_2(2/3t) + S_3(2t/3)\} \\ + S_2(t) + S_3(t). \end{aligned} \tag{4.43}$$

We can increase the number of moving nodes by both variables. Due to the difficulty of implementing this procedure, we will present this routine work to symbolic mathematics packages, in particular, we will use Maple tools.

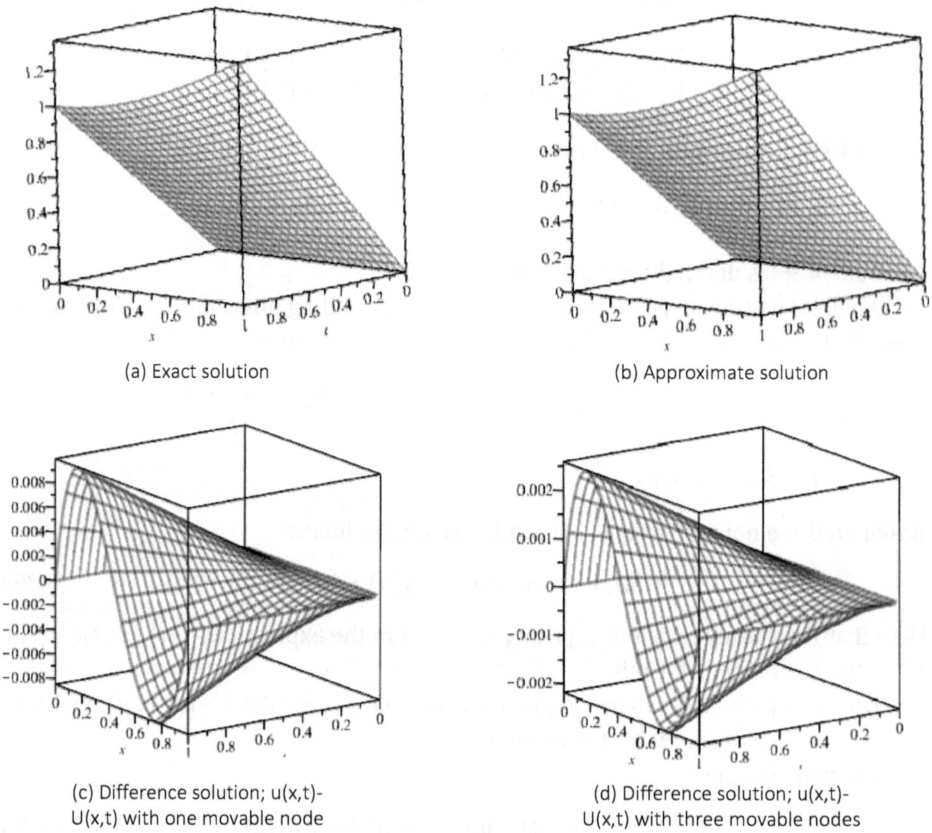

(a) Exact solution

(b) Approximate solution

(c) Difference solution; u(x,t)-U(x,t) with one movable node

(d) Difference solution; u(x,t)-U(x,t) with three movable nodes along x

Fig. 4.22: Comparison of problem (4.1).

Note that, for convenience, the approach was presented for a simple parabolic eq. (4.1). The sequence of presentation and the method for obtaining an approximate solution do not change if, instead of eq. (4.1), we consider a more general form.

4.5.2 Numerical experiments

We will study several examples in which the exact solution is known.

Problem **1.** Consider problems (4.1)–(4.3) with the following input parameters: $\sigma = 1, b = 1,$

$$T = 1,\ f(x,t) = \exp(-x)(1-t) + x,\ \varphi(x) - 0,\ u_0(t) = t,\ u_b(t) = t/e + t.$$

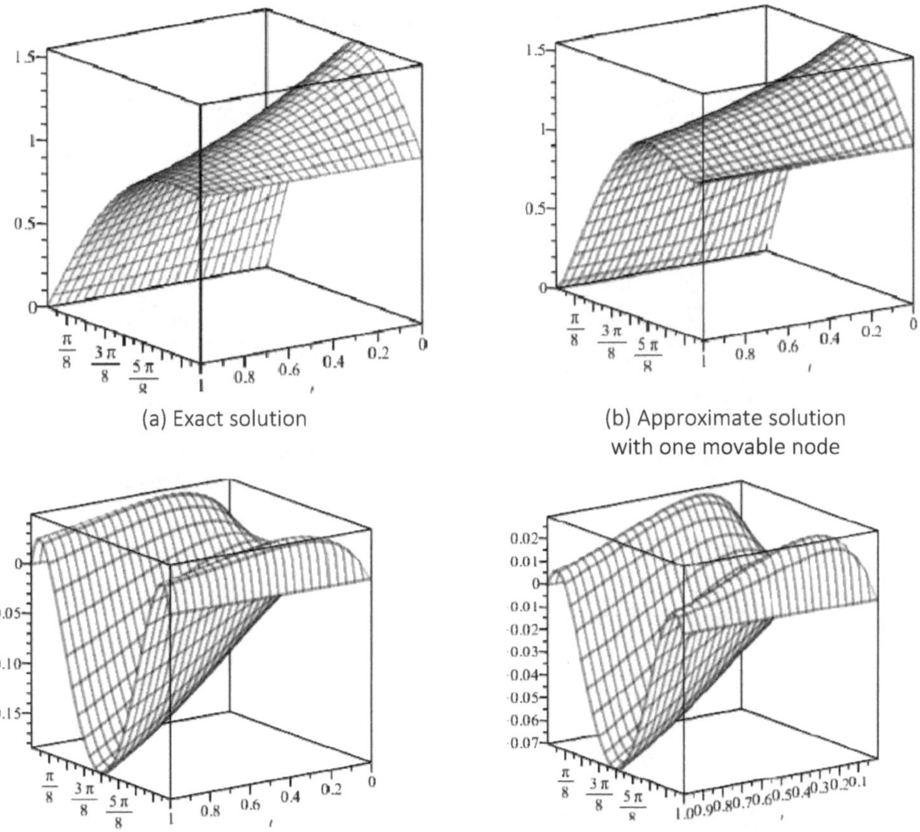

(a) Exact solution

(b) Approximate solution with one movable node

(c) Difference solution; u(x,t)-U(x,t) with one movable node

(d) Difference solution; u(x,t)-U(x,t) with three movable nodes along x and along t

Fig. 4.23: Comparison of results of problem (4.2).

Exact solution to the problem

$$u(x,t) = \exp(-x)t + tx.$$

Figure 4.22(a) and (b) shows a comparison of the exact and approximate solutions obtained using one movable node. It is very difficult to distinguish visually, so Fig. 4.22 (c) shows the difference between the exact and approximate solutions, showing a good result. When used with three movable nodes along the variable x, the result improves (the difference between the solution is shown in Fig. 4.22(d)).

Problem **2.** Let the parameters of problems (4.1)–(4.3) have the form: $\sigma = 1$, $b = \pi$, $T = 1$:

$$T = 1, \ f(x,t) = 0, \quad \varphi(x) = \sin x + x/\pi, \quad u_0(t) = 0, \ u_b(t) = 1.$$

Exact solution of the problem

$$u(x,t) = \exp(-t)\sin(x) + x/\pi.$$

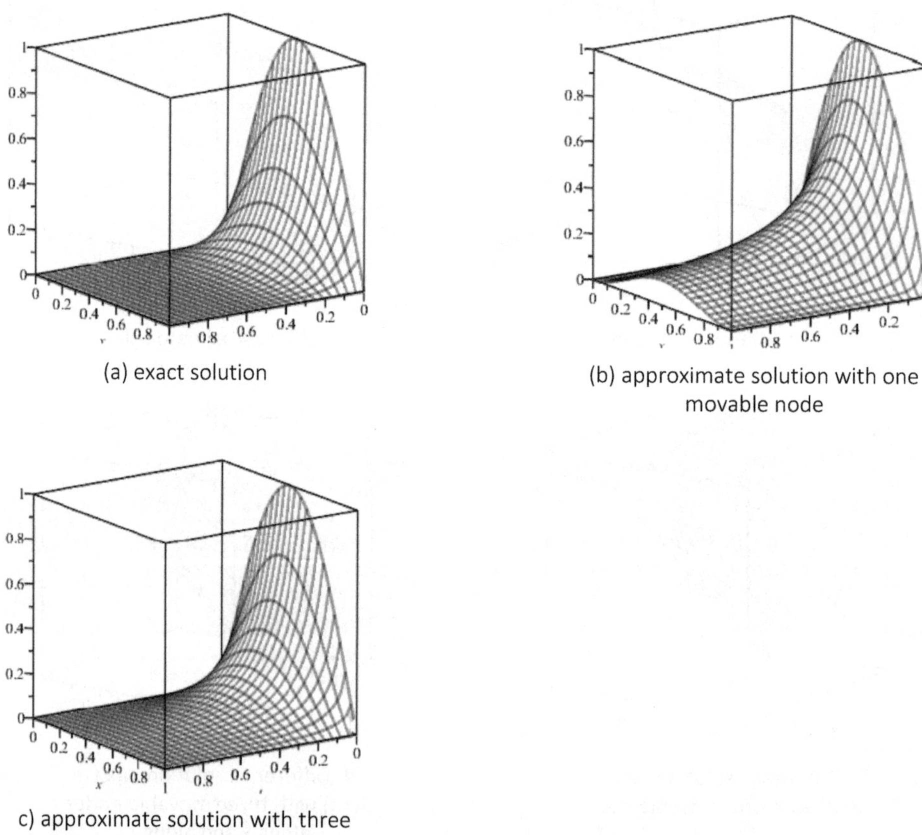

(a) exact solution

(b) approximate solution with one movable node

c) approximate solution with three movable nodes in x and t

Fig. 4.24: Exact and approximate solutions to problem (4.3) with different movable nodes.

Figure 4.23 reflects a comparative analysis of the approximate and exact solutions for problem (4.2). The exact solution of problem (4.2) is shown in Fig. 4.23(a), and the approximate solution with one movable node is shown in Fig. 4.23(b). Comparison of Fig. 4.23(a) and (b) shows qualitative agreement between the obtained solutions. The difference between the exact and approximate solutions with a single-point moving node is shown in Fig. 4.23(c). To refine the approximate solution, additional alternating nodes were used. When using three moving nodes along the x variable and along the t variable, the difference between the exact and approximate solutions was reduced by two times compared to one moving node (Fig. 4.23(d)).

***Problem* 3.** Now consider problems (4.1)–(4.3) with the input parameters: $\sigma=1$, $b=1$, $T=1$,

$$f(x,t)=0, \quad \varphi(x)=\sin(\pi x), \quad u_0(t)=0, \quad u_b(t)=0.$$

Exact solution to the problem

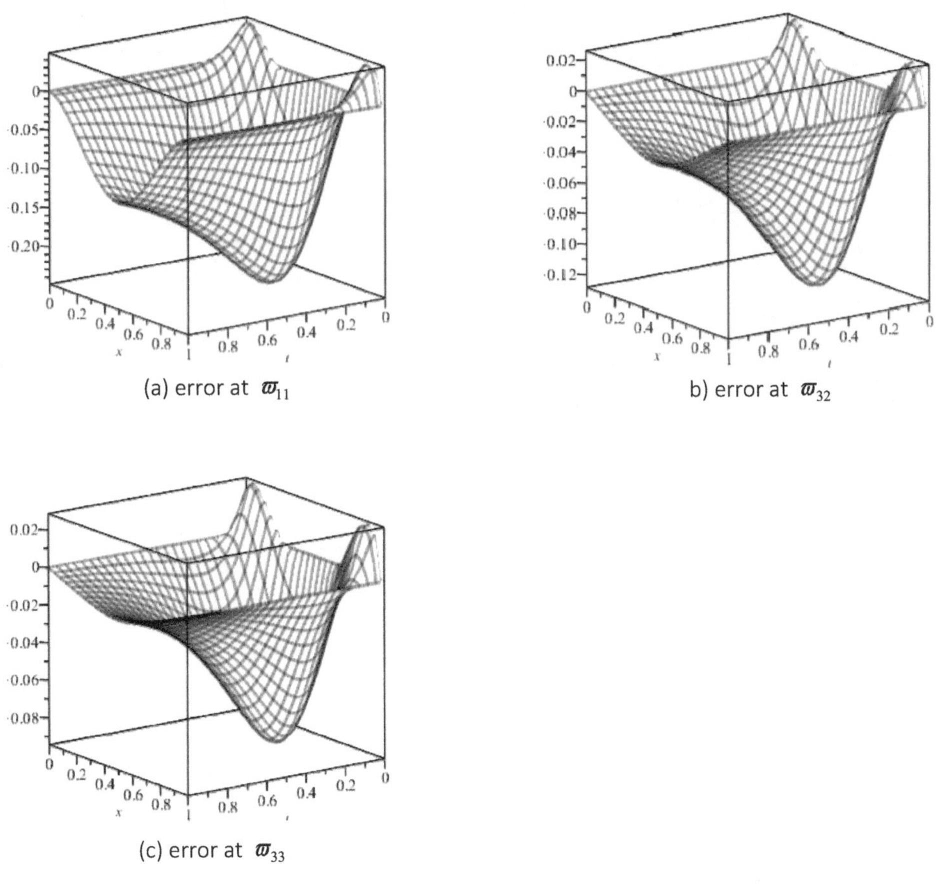

(a) error at ϖ_{11}

b) error at ϖ_{32}

(c) error at ϖ_{33}

Fig. 4.25: Errors in solving problem (4.3) with different movable nodes.

$$u(x,t) = \exp(-\pi^2 t)\sin(\pi x).$$

The solution to this problem with one moving node is

$$U(x,t) = \frac{x(1-x)\sin(\pi x)}{2t + x(1-x)}.$$

In this problem, one can also find a qualitative coincidence between the exact and approximate solutions with one moving node (Fig.4.24). To refine it, we increase the number of moving nodes.

Figure 4.25 shows graphs showing errors (the difference between the exact and approximate solutions) depending on the number of movable nodes.

Increasing the number of movable nodes will lead to a decrease in the error of the solutions obtained. In this case, the decrease in error, in this problem, is more noticeable with an increase in the number of movable nodes in the variable t.

Chapter 5
Solving boundary value problems for elliptic equations

This chapter is devoted to the application of the movable node method to linear elliptic problems of mathematical physics. Using a finite-difference movable node, by analogy with the previous chapters, we obtain an approximate analytical solution. In specific examples, a method for obtaining an analytical solution for various applied problems is given. A method for using movable nodes for Dirichlet and Neumann boundary value problems is described. Using partial approximation of partial derivatives, a method for improving the obtained approximate solution is described.

5.1 Solving the Dirichlet problem for two-dimensional boundary value problems by the method of movable nodes

Most problems in physics and engineering cannot be solved without linear and nonlinear differential equations in partial derivatives of an elliptic equation. The most optimal and universal method for solving them is the finite-difference method or the grid method.

When studying stationary processes of various physical nature (oscillations, thermal conductivity, diffusion, hydrodynamics, etc.), one usually comes to elliptic equations. The most common equation of this type is the Poisson equation [29].

Now let us consider the application of the moving node method to two-dimensional boundary value problems to obtain an approximate solution to the differential equation.

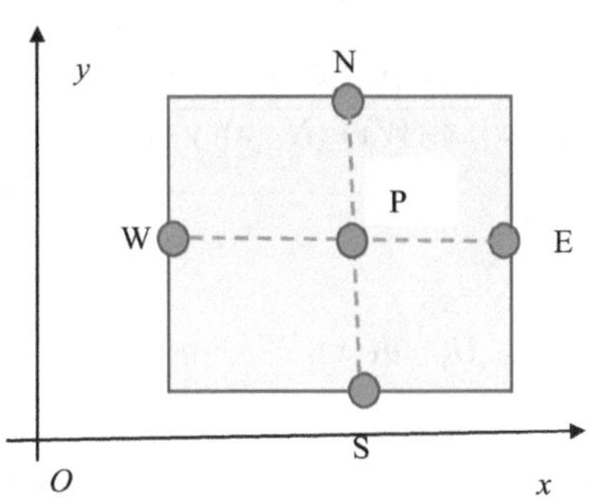

Fig. 5.1: Solution domain.

Let us consider the two-dimensional Poisson equation in a rectangle $\Omega = \{(x,y) | W \le x \le E,\ S \le y \le N\}$ (Fig. 5.1):

https://doi.org/10.1515/9783111308173-006

$$\Delta u(x,y)=\frac{\partial^2 u}{\partial x^2}+\frac{\partial^2 u}{\partial y^2}=f(x,y), \tag{5.1}$$

with Dirichlet boundary conditions:

$$u(W,y)=u_W(y),\ u(E,y)=u_E(y),\ u(x,S)=u_S(x),\ u(x,N)=u_N(x). \tag{5.2}$$

We assume that the input parameters of problems (5.1) and (5.2) are smooth and there is a unique solution. Take an arbitrary point in a rectangle $(x,y)\in\Omega$. Approximate the second-order partial derivatives with a movable difference ratio:

$$\frac{\partial^2 u}{\partial x^2}\approx\frac{2}{E-W}\left[\frac{u_E(y)-u(x,y)}{E-x}+\frac{u(x,y)-u_W(y)}{x-W}\right], \tag{5.3}$$

$$\frac{\partial^2 u}{\partial y^2}\approx\frac{2}{N-S}\left[\frac{u_N(x)-u(x,y)}{N-y}+\frac{u(x,y)-u_S(x)}{y-S}\right]. \tag{5.4}$$

Substituting eqs. (5.3) and (5.4) into eq. (5.1), we obtain an approximate algebraic equation:

$$\begin{aligned}&\frac{2}{E-W}\left[\frac{u_E(y)-u(x,y)}{E-x}+\frac{u(x,y)-u_W(y)}{x-W}\right]\\&+\frac{2}{N-S}\left[\frac{u_N(x)-u(x,y)}{N-x}y+\frac{u(x,y)-u_S(x)}{x-S}\right]\approx f(x,y),\end{aligned} \tag{5.5}$$

Denote by $U(x,y)$ the approximate solution obtained from eq. (5.5) by replacing the approximate sign with an exact equality, as a result we obtain:

$$\begin{aligned}U(x,y)=&\frac{1}{(E-x)(x-W)+(N-y)(y-S)}\cdot\left[\frac{(N-y)(y-S)}{E-W}((x-W)U_E+(E-x)U_W)\right.\\&\left.+\frac{(E-x)(x-W)}{N-S}((y-S)U_N+(N-y)U_S)\right]+\frac{(E-x)(x-W)(N-y)(y-S)}{2((E-x)(x-W)+(N-y)(y-S))}f(x,y).\end{aligned} \tag{5.6}$$

In the approximate solution of problem (5.6), it is set

$$U_E=u_E(y),\quad U_W=u_W(y),\quad U_N=u_N(x),\quad U_S=u_S(x).$$

Formula (5.6) is an approximate analytical solution to the Dirichlet problem of the Poisson equation in a rectangle, obtained using a movable node. The approximate solution (5.6) gives an acceptable solution in many practical problems. Let us consider examples.

Note that the formula for the approximate solution (5.6) is obtained for a rectangular region. If the region is not rectangular, it is easy to obtain a similar formula of type (5.6). But in this case, the coordinate system for the region should be chosen so that the lines parallel to the coordinate axes intersect the region boundary only at two points. For example, the convexity of the region is a sufficient condition for this condition to be satisfied. Let the point P be an arbitrary point with coordinates $(x,y) \in \Omega$ located inside the region. If P changes its position inside the region, the boundary points W, E, S, N change their positions while on the region boundary. The point $P(x,y)$ is considered a movable node.

5.1.1 Test examples

1. Consider the Laplace equation in a rectangle $[0,1] \times [0,1]$

$$\frac{\partial^2 u}{\partial x^2} + \frac{\partial^2 u}{\partial y^2} = 0,$$

with boundary conditions $u(0,y) = u_W(y) = 0$, $u(1,y) = u_E(y) = 0$, $u(x,0) = u_S(x) = 0, u(x,1) = u_N(x) = x$. The exact solution of this problem is $u(x,y) = xy$. Using the approximate solution according to formula (5.6), we obtain

$$U(x,y) = \frac{1}{(1-x)x + (1-y)y} \cdot \left[\frac{(1-y)y}{1}((xy + (1-x)0) + \frac{(1-x)x}{1}(yx + (1-y)0)\right].$$

From here it is easy to obtain that $U(x,y) = xy$. This shows that the approximate solution coincides with the exact solution.

2. The function $u(x,y) = x^2 - y^2$, with boundary conditions $u(0,y) = u_E(y) = -y^2$, $u(1,y) = u_E(y) = 1 - y^2$, $u(x,0) = u_S(x) = x^2$, $u(x,1) = u_N(x) = x^2 - 1$ satisfies the Laplace equation. Relation (5.6) gives us an identical result.

3. The function

$$u(x,y) = \ln(x^2 + y^2), \tag{5.7}$$

with boundary conditions

$$u(1,y) = u_W(y) = \ln(1 - y^2), \quad u(2,y) = u_E(y) = \ln(4 + y^2),$$
$$u(x,0) = u_S(x) = \ln(x^2), \quad u(x,1) = u_N(x) = \ln(x^2 + 1)$$

in the domain $[1,2] \times [0,1]$ satisfies the Laplace equation. An approximate solution based on eq. (5.6) yields

$$u(x,y) = \frac{y(1-y)((x-1)\ln(4+y^2) + (2-x)\ln(1+y^2)))}{(2-x)(x-1) + y(1-y)} + \frac{(2-x)(x-1)(y\ln(1+x^2) + (1-y)\ln(x^2))}{(2-x)(x-1) + y(1-y)}. \tag{5.8}$$

We compare the obtained solution with the exact solution. This comparison is made in Fig. 5.2. Comparison of Fig. 5.2(a) and (b) shows the closeness of the exact and approximate solutions. A more detailed comparison is given in Fig. 5.2(c) and (d). Figure 5.2(c) shows the graphs for the cross sections $y = 0.1$ and 0.7, the solid lines correspond to the exact solution and the dotted lines to the approximate one. Figure 5.2(d) shows the difference of the exact and approximate solutions obtained for the cross sections $y = 0.1$ and 0.7 (the solid curve corresponds to the cross section and the dotted line to the cross section $y = 0.7$).

Thus, formula (5.8) is a good approximation for the exact solution.

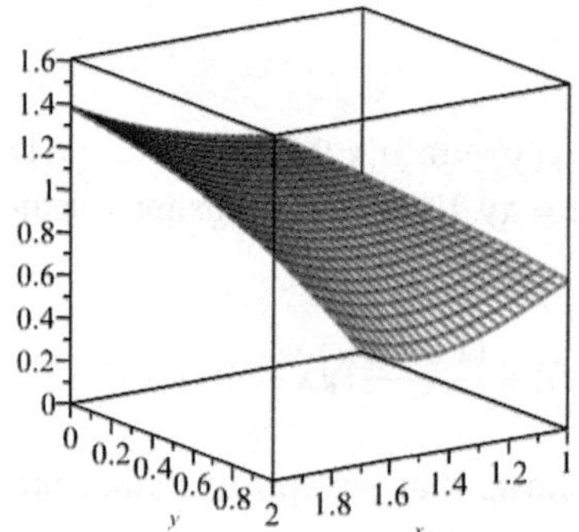

(a) Exact solution

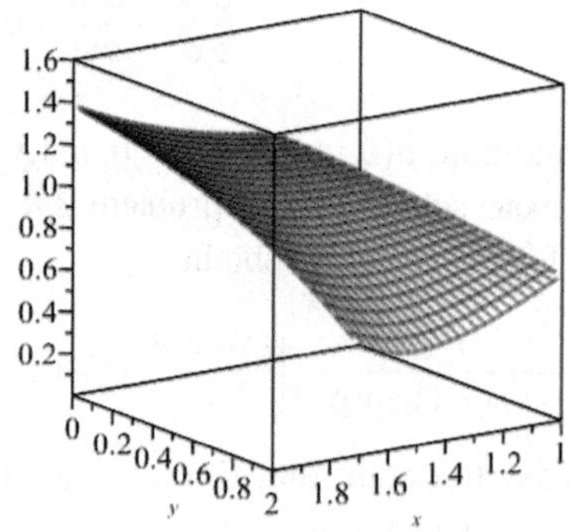

(b) Approximate solution

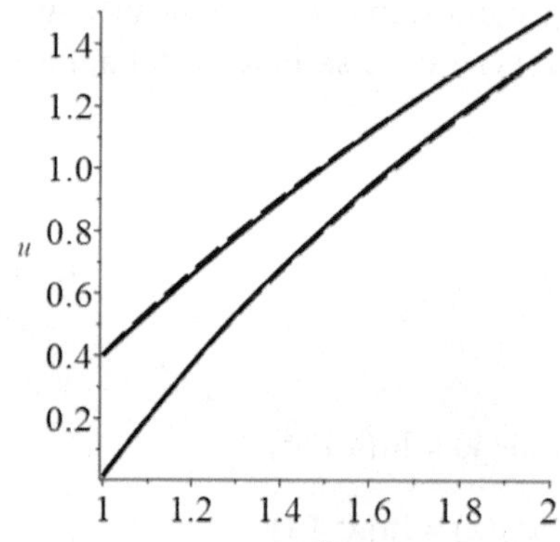

(c) Comparisons of the solution for different cross-sections

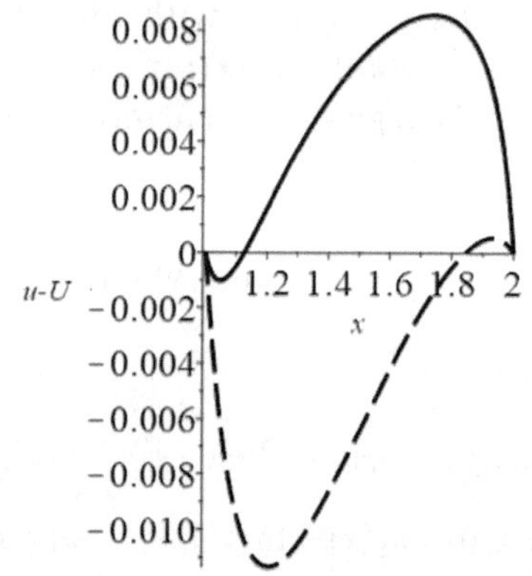

(d) Difference between the exact and approximate solutions

Fig. 5.2: Comparison of the solution.

4. The function $U(x,y) = x^3 - y^3$, with boundary conditions $U(0,y) = -y^3$, $U(1,y) = 1 - y^3$, $U(x,0) = x^3$, $U(x,1) = x^3 - 1$ satisfies eq. (5.1) for $f(x,y) = 6x - 6y$. Based on eq. (5.6), the approximate solution has the form:

$$u(x,y) = \frac{xy(y-x) + y^4(1-y) + x^4(x-1)}{y(y-1) + x(x-1)}.$$

The maximum absolute difference between the exact and approximate solutions, calculated by points $x_i = 1 + ih,\ y_j = jh,\quad i,j = 1,2,\ldots,n,\ h = 0.1$, is 0.048.

5.1.2 Flow in an ellipsoidal pipe

The equation describing the one-dimensional flow of a viscous fluid in an ellipsoidal pipe has the form [5]:

$$\frac{\partial^2 u}{\partial y^2} + \frac{\partial^2 u}{\partial z^2} = -\frac{\Delta p}{\mu l}. \tag{5.9}$$

Here u is the flow velocity, μ is the flow viscosity, $\Delta p/l$ ($\Delta p/l = \text{const}$) is the pressure drop. Equation (5.9) is considered in the region $\frac{y^2}{a^2} + \frac{z^2}{b^2} \le 1$ (the cross section of an ellipsoidal pipe, Fig. 5.3), the boundary condition is the no-slip condition ($U = 0$).

We replace eq. (5.1) with a difference:

$$\frac{2}{y_E - y_W}\left(\frac{U_E - U}{y_E - y} - \frac{U - U_W}{y - y_W}\right) + \frac{2}{z_N - z_S}\left(\frac{U_N - U}{z_N - z} - \frac{U - U_S}{z - z_S}\right) = -\frac{\Delta p}{\mu l}.$$

Hence, taking into account that

$$z_N = b\sqrt{1 - y^2/a^2},\ z_S = -b\sqrt{1 - y^2/a^2},$$
$$y_E = a\sqrt{1 - z^2/b^2},\ y_W = -a\sqrt{1 - z^2/b^2},$$

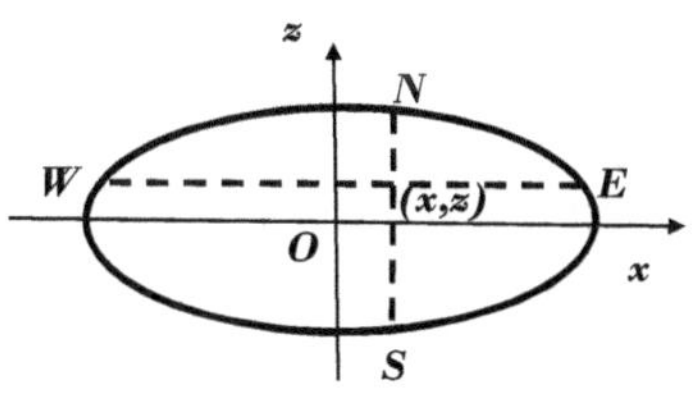

Fig. 5.3: Ellipsoidal pipe cross section.

we obtain

$$U = \frac{a^2 b^2}{2(a^2+b^2)}\left(1 - \frac{y^2}{a^2} - \frac{z^2}{b^2}\right)\frac{\Delta p}{\mu l},$$

which coincides with the exact solution [22].

5.1.3 Flow in a rectangular pipe

Equation (5.1) also describes the flow of an incompressible viscous fluid in a rectangular pipe [5]. Let us denote the height of the rectangle parallel to the axis Oz by $2h$ and the base parallel to the axis Oy by $2\sigma h$, where σ is any positive constant. We draw the axis through the center of the rectangle and direct it downstream.

Transform eq. (5.1) to a dimensionless form. We take the height, h, as the length scale, and the value $h^2/\mu \cdot \Delta p/l$, as the velocity scale. We introduce the following dimensionless quantities:

$$Y = y/h, \quad Z = z/h, \quad V = U\mu\, l/(h^2 \Delta p).$$

Substituting into eq. (5.1), we obtain

$$\frac{\partial^2 V}{\partial Y^2} + \frac{\partial^2 V}{\partial Z^2} = -1. \tag{5.10}$$

Boundary conditions for eq. (5.10)

$$V(Y,-1) = 0, \quad V(Y,1) = 0, \quad V(-\sigma, Z) = 0, \quad V(\sigma, Z) = 0. \tag{5.11}$$

We replace eq. (5.10) with a difference equation, and taking into account the boundary condition (5.11), we have

$$\frac{2}{2\sigma}\left(\frac{-V}{\sigma - Y} - \frac{V}{Y+\sigma}\right) + \frac{2}{1+1}\left(\frac{-V}{1-Z} - \frac{V}{Z+1}\right) = -1.$$

From here we determine the approximate analytical solution:

$$V = \frac{1}{2}\frac{(\sigma^2 - Y^2)(1-Z^2)}{1 - Z^2 + \sigma^2 - y^2}. \tag{5.12}$$

The exact solution to the problem is

$$u = \frac{16\sigma^2}{\pi^3}\sum_{n=0}^{\infty}\frac{(-1)^n}{(2n+1)^3}\left[1 - \frac{ch\left(\frac{2n+1}{2}\frac{\pi}{\sigma}Y\right)}{ch\left(\frac{2n+1}{2}\frac{\pi}{\sigma}\right)}\right]\cos\left(\frac{2n+1}{2}\frac{\pi}{\sigma}Z\right).$$

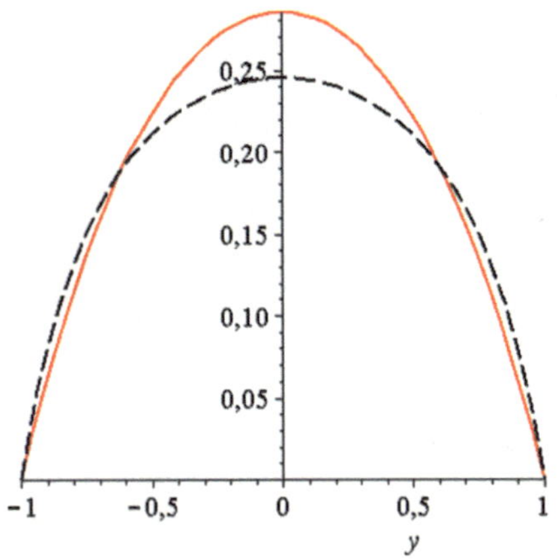

Fig. 5.4: Comparison of the solution by formula (5.12).

Figure 5.4 shows a comparison of the exact and approximate solutions on the section $x = 0$ at $\sigma = 1$. The maximum absolute difference between the exact and approximate solutions is 0.045.

5.1.4 Calculating the temperature field in a solid by a movable control volume

The analytical approximation described in the previous paragraphs was obtained on the basis of a movable separation ratio. This paragraph describes obtaining an approximate analytical solution based on a movable control volume. The well-known control volume method used in numerical analysis is used taking into account its movability. The movability method makes it possible to obtain an analytical representation of the solution to the problem under consideration. In this case, obtaining an analytical solution method is achieved by averaging the equation describing the flows over the control volume.

Let us consider the problem of temperature propagation on a homogeneous body. The problem is reduced to solving the Laplace equation with boundary conditions [1] (we will choose the Cartesian coordinates as shown in Fig. 5.5):

$$T(0,y) = 0, \ T(1,y) = 0,$$
$$T(x,0) = T_0, \ T(x,1) = 0.$$

Exact solution to the problem

$$T(x,y) = \sum_{n=1}^{\infty} A_n \sin(n\pi x) sh[n\pi(y-1)],$$

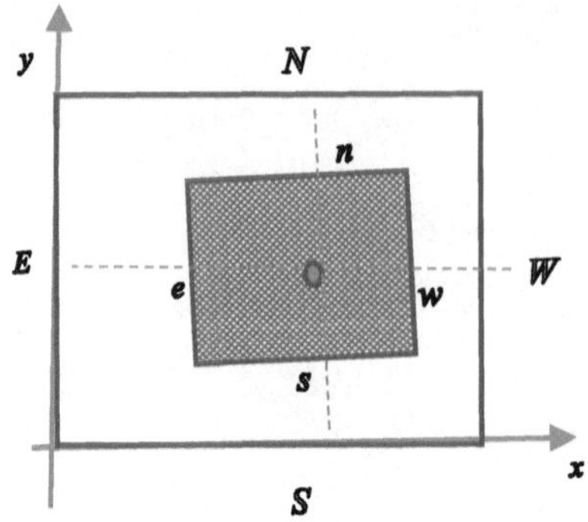

Fig. 5.5: Movable control volume.

where

$$A_n = \frac{2T_0}{n\pi}\frac{[(-1)^n - 1]}{sh(n\pi)}.$$

We average the Laplace equation over the control volume (shaded area) [9]. Here we consider the control volume to be movable:

$$\begin{aligned}\iint\limits_D \left(\frac{\partial^2 T}{\partial x^2} + \frac{\partial^2 T}{\partial y^2}\right) dxdy &= \iint\limits_D \frac{\partial^2 T}{\partial x^2} dxdy + \iint\limits_D \frac{\partial^2 T}{\partial y^2} dxdy \\ &= \int\limits_{y/2}^{(1+y)/2} dy \int\limits_{x/2}^{(1+x)/2} \frac{\partial^2 T}{\partial x^2} dx + \int\limits_{x/2}^{(1+x)/2} dx \int\limits_{y/2}^{(1+y)/2} \frac{\partial^2 T}{\partial y^2} dy.\end{aligned} \tag{5.13}$$

Since

$$\int\limits_{y/2}^{(1+y)/2} dy \int\limits_{x/2}^{(1+x)/2} \frac{\partial^2 T}{\partial x^2} dx = \int\limits_{y/2}^{(1+y)/2} \left(\left.\frac{\partial T}{\partial x}\right|_{(1+x)/2} - \left.\frac{\partial T}{\partial x}\right|_{x/2}\right) dy \approx \frac{1}{2}\left(\frac{T_E - T}{1-x} - \frac{T - T_W}{x}\right). \tag{5.14}$$

The last expression is obtained under the condition that the derivatives on the faces of the control volume are considered constant and the derivative is replaced by the central difference. In eq. (5.14), the value of the temperature at the boundary point W, the designation has a similar meaning. Proceeding similarly with the second expression in eq. (5.13), we obtain

$$\iint\limits_D \left(\frac{\partial^2 T}{\partial x^2} + \frac{\partial^2 T}{\partial y^2}\right) dxdy \approx \frac{1}{2}\left(\frac{T_E - T}{1-x} - \frac{T - T_W}{x}\right) + \frac{1}{2}\left(\frac{T_N - T}{1-y} - \frac{T - T_S}{y}\right).$$

Hence, using the boundary conditions, instead of the Laplace equation, we obtain

$$\left(\frac{-T}{1-x}-\frac{T}{x}\right)+\left(\frac{-T}{1-y}-\frac{T-T_0}{y}\right)=0.$$

Thus, we have an approximate analytical solution

$$T(x,y)=\frac{(1-x)x(1-y)T_0}{x(1-x)+y(1-y)}.$$

The maximum absolute difference between the exact and approximate solutions, calculated by points $x_i=ih,\ y_j=jh,\quad i,j=1,2,\ldots n,\ h=0.01$, is 0.015.

Thus, the method presented here allows obtaining a solution to Dirichlet problems.

5.1.5 Obtaining a solution for flow in a porous pipe using the moving control volume method

Consider the flow of a viscous Newtonian fluid through pipes filled with a porous medium. We assume the flow to be steady and one-dimensional, then from the Rakhmatulin equation we obtain the equation [10, 24] in the Cartesian coordinate system:

$$\frac{\partial}{\partial x}\left(f\frac{\partial u}{\partial x}\right)+\frac{\partial}{\partial y}\left(f\frac{\partial u}{\partial y}\right)=-f\frac{\Delta p}{\mu l}+Ku. \tag{5.15}$$

Here $u(x,y)$ is the flow velocity, the Oz-axis is directed along the pipe axis, $\Delta P/l$ is the pressure drop, μ is the viscosity of the fluid, f is the porosity of the medium, $K=\frac{a}{d^2}\left(\frac{1-f}{f}\right)$ is the interaction coefficient obtained based on the Kozeny-Karman relation and d is the characteristic size of the porous medium (we assume the pressure drop and viscosity of the fluid to be constant). We assume the boundary of the pipe to satisfy the equation:

$$\frac{x^n}{a^n}+\frac{y^n}{b^n}=1. \tag{5.16}$$

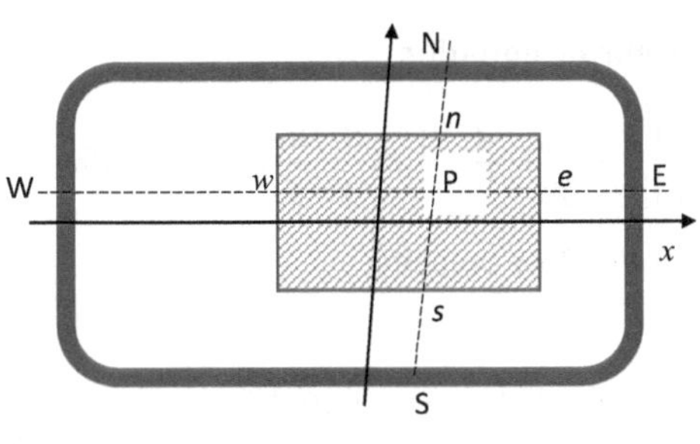

Fig. 5.6: Pipe cross section.

The pipe cross section is shown in Fig. 5.6. In the case of $n=2$ and $a=b$ (eq. (5.16)) represents the cross section of a round pipe with a radius a; for $n=2,\ a\neq b$ we obtain the cross section of an ellipsoidal pipe with semiaxes a and b. For sufficiently large even n, the cross section of the pipe approaches a rectangle.

Thus, find a solution to eq. (5.15) with no-slip conditions at the pipe boundaries using the moving volume method.

Let $P(x,y)$ be an arbitrary point inside the pipe cross section (Fig. 5.6). We draw parallel lines through this point along the coordinate axes until they intersect with the pipe boundary. We denote these points by E, W, N, and S. The coordinates of these boundary points are $E(x_E,y_E)$, $W(x_W,y_W)$, $N(x_N,y_N)$ and $S(x_S,y_S)$. We choose a rectangular control volume as follows. We choose the midpoint of points P and E ($e=0.5(E+P)$) W and P ($w=0.5(W+P)$); similarly $n=0.5(N+P)$ and $s=0.5(S+P)$. Let us denote the coordinates of the points $e(x_e,y_e)$, $w(x_w,y_w)$, $n(x_n,y_n)$ and $s(x_s,y_s)$ (by drawing parallel lines through these points along the coordinate axes we obtain a rectangle (the shaded area in Fig. 5.6)). When P changes its position, the control volume also changes. We integrate eq. (5.15) over the control volume:

$$\iint\limits_D \left[\frac{\partial}{\partial x}\left(f\frac{\partial u}{\partial x}\right)+\frac{\partial}{\partial y}\left(f\frac{\partial u}{\partial y}\right)\right]dxdy=\iint\limits_D [fA+Ku]dxdy. \tag{5.17}$$

Here D: $(x_w \le x \le x_e;\ y_s \le y \le y_n)$, $A=-\Delta P/(\mu l)$. We transform the first term of eq. (5.17) as follows:

$$\iint\limits_D \frac{\partial}{\partial x}\left(f\frac{\partial u}{\partial x}\right)dxdy=\int\limits_{y_s}^{y_n} dy\int\limits_{x_w}^{x_e}\frac{\partial}{\partial x}\left(f\frac{\partial u}{\partial x}\right)dx=\frac{y_N-y_S}{2}\left[f\frac{\partial u}{\partial x}\bigg|_{x_e}-f\frac{\partial u}{\partial x}\bigg|_{x_w}\right]$$
$$\approx\frac{y_N-y_S}{2}\left(f_{x_e}\frac{u_E-u}{x_E-x}-f_{x_w}\frac{u-u_W}{x-x_W}\right).$$

The second term of the left-hand side is approximated similarly. The integral on the right-hand side of eq. (5.17) is replaced by the expression:

$$\iint\limits_D [fA+Ku]dxdy\approx(fA+Ku)\frac{x_E-x_W}{2}\cdot\frac{y_N-y_S}{2}.$$

Thus, eq. (5.15) is replaced by the following algebraic equation:

$$\frac{2}{x_E-x_W}\left[f(e)\frac{u_E-u}{x_E-x}-f(w)\frac{u-u_W}{x-x_W}\right]+\frac{2}{y_N-y_S}\left[f(n)\frac{u_N-u}{y_N-y}-f(s)\frac{u-u_W}{y-y_S}\right]=(fA+Ku).$$

Using the boundary, no-slip condition, we find the flow velocity:

$$u=\frac{2G\cdot H}{2[f(e)(x-x_W)+f(w)(x_E-x)]G+[f(n)(y-y_S)+f(s)(y_N-y)]H+K\cdot G\cdot H}\times\frac{f(P)\cdot A}{2}, \tag{5.18}$$

where the notation

$$H = (x_E - x_W)(x_E - x)(x - x_W), \quad G = (y_N - y_S)(y_N - y)(y - y_S).$$

Equation (5.18) represents the flow velocity distribution for pipes filled with a porous medium of different cross sections.

In connection with the fact that the points E, W, N, and S are on the boundary of eq. (5.16), we obtain

$$y_N = b\sqrt[n]{1 - \frac{x^n}{a^n}}, \quad y_S = -b\sqrt[n]{1 - \frac{x^n}{a^n}}, \quad x_E = a\sqrt[n]{1 - \frac{y^n}{b^n}}, \quad x_W = a\sqrt[n]{1 - \frac{y^n}{b^n}}. \tag{5.19}$$

If in eq. (5.18) we set $f \equiv 1$. we obtain a solution to the problem in the absence of a porous layer:

$$u = \frac{a^2 b^2 \left[\sqrt[n]{\left(1 - \frac{y^n}{b^n}\right)^2} - \frac{x^2}{a^2}\right] \cdot \left[\sqrt[n]{\left(1 - \frac{x^n}{a^n}\right)^2} - \frac{y^2}{b^2}\right]}{a^2 \left[\sqrt[n]{\left(1 - \frac{y^n}{b^n}\right)^2} - \frac{x^2}{a^2}\right] + b^2 \left[\sqrt[n]{\left(1 - \frac{x^n}{a^n}\right)^2} - \frac{y^2}{b^2}\right]} \cdot \frac{A}{2}. \tag{5.20}$$

Let us consider special cases. Let us set in eq. (5.20) $n = 2$ and $a = b$, that is, we consider the flow in a round pipe with radius a. Then we have

$$u = \frac{a^2}{4}\left(1 - \frac{r^2}{a^2}\right) \cdot \frac{A}{2}. \tag{5.21}$$

Here $r^2 = x^2 + y^2$. Thus, we obtain Poiseuille's law for a round pipe. If $n = 2$ and $a \neq b$ then we have a flow in an ellipsoidal pipe. In this case, from eq. (5.20)

$$u = \frac{a^2 b^2}{2(a^2 + b^2)}\left(1 - \frac{x^2}{a^2} - \frac{y^2}{b^2}\right) \cdot \frac{A}{2}. \tag{5.22}$$

The law of flow velocity distribution according to formula (5.22) coincides with the exact solution [21].

To obtain an approximate analytical solution for a pipe with a rectangular cross section, we pass to the limit in eq. (5.20) at. After simple operations we obtain

$$u = \frac{a^2 b^2 \left(1 - \frac{x^2}{a^2}\right)\left(1 - \frac{y^2}{b^2}\right)}{a^2 \left(1 - \frac{x^2}{a^2}\right) + b^2 \left(1 - \frac{y^2}{b^2}\right)} \cdot \frac{A}{2}. \tag{5.23}$$

However, formula (5.23) does not coincide with the exact solution [21], but comes close to it. To obtain a solution for a flat pipe, we pass to the limit in eq. (5.23) at $a \to \infty$, then we have

$$u = b^2\left(1 - \frac{y^2}{b^2}\right)\cdot\frac{A}{2}. \tag{5.24}$$

Solution (5.24) accurately reflects the solution to the problem for a flat pipe with height $2b$ [21].

Let us compare the exact solution for a rectangular pipe, which has the form [33]:

$$u = A\frac{16a^2}{\pi^3}\sum_{n=0}^{\infty}\frac{(-1)^n}{(2n+1)^3}\left[1 - \frac{ch\left(\frac{2n+1}{2}\frac{\pi}{a}y\right)}{ch\left(\frac{2n+1}{2}\frac{\pi b}{a}\right)}\right]\cos\left(\frac{2n+1}{2}\frac{\pi}{a}x\right) \tag{5.25}$$

with the approximate solution (5.23) (b is the height of the rectangle, the parallel axis Oy, the base of the parallel axis Ox and the width $2a$; the axis Oz passes through the center of the rectangle and is directed downstream). Figure 5.7 shows the exact and approximate solutions for on the section $x = 1$ ($A = 1,\ b = 1,\ a = 2.5$).

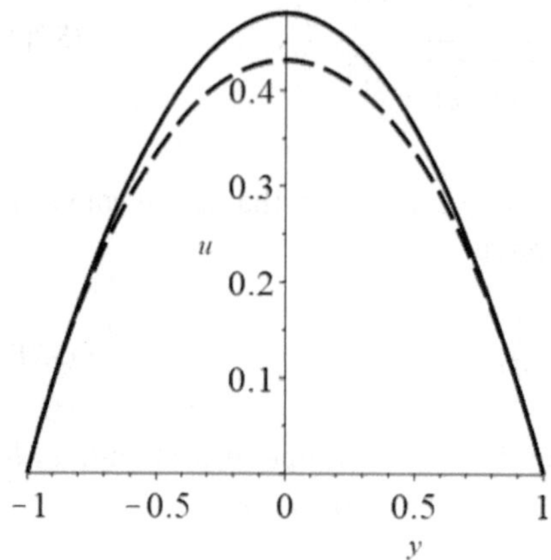

Fig. 5.7: Comparison of the solution.

Thus, formula (5.20) describes well the velocity distribution for pipes of type (5.16) with different cross sections.

5.2 Solution of the Neumann problem for two-dimensional boundary value problems by the MNM

Section 5.1 considers examples of obtaining approximate solutions to various Dirichlet problems. Now consider problems in which the boundary conditions contain Neumann-type conditions.

5.2.1 Flow in a pipe with a triangular cross section

Let us consider the flow in a pipe with an equilateral triangular cross section (Fig. 5.8).

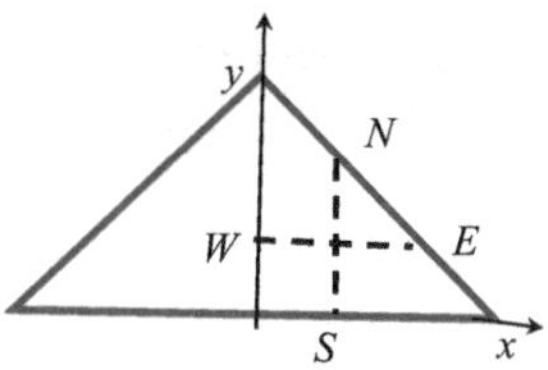

Fig. 5.8: Section of a triangular pipe.

Mathematically, this problem is reduced to solving the Poisson equation (5.1) with no-slip conditions at the pipe boundaries.

We take the coordinate system as shown in Fig. 5.8. Since the flow is symmetrical relative to the *Oy* axis, we consider only the right side of the triangle. Then the flow process is described by eq. (5.1) with no-slip boundary conditions at the triangle boundaries and the symmetry conditions hold on the axis of symmetry:

$$\frac{\partial u}{\partial x} = 0. \tag{5.26}$$

For this case, the difference equation can be rewritten as follows: $u_E = 0$, $u_N = 0$, $u_S = 0$, $x_W = 0$, $y_S = 0$:

$$\frac{2}{x_E}\left[\frac{-u}{x_E - x} - \frac{u - u_0}{x}\right] + \frac{2}{y_N}\left[\frac{-u}{y_N - y} - \frac{u}{y}\right] = -A. \tag{5.27}$$

Here u_0 is the value of the velocity on the axis of symmetry. In eq. (5.27) we pass to the limit at $x \to 0$ then $\lim\limits_{x\to 0}\frac{u-u_0}{x} = \frac{\partial u}{\partial x}$ and eq. (5.27) can be rewritten as follows:

$$\frac{2}{x_E}\left[\frac{-u}{x_E} - \frac{\partial u}{\partial x}\right] + \frac{2}{y_N}\left[\frac{-u_0}{y_N - y} - \frac{u_0}{y}\right] = -A.$$

From here, taking into account condition (5.26), we obtain

$$u_0 = \frac{A}{2} \cdot \frac{y(y_N - y)x_N^2}{y(y_N - y) + x_E^2}. \tag{5.28}$$

Thus, we determined the velocity on the axis of symmetry. From eq. (5.27) we find the velocity taking into account (5.28):

$$u_0 = \frac{A}{2} \cdot \frac{(x_E - x)y(y_N - y)}{x(x_E - x) + y(y_N - y)}\left[x + \frac{y(y_N - y)x_E}{y(y_N - y) + x_E^2}\right]. \tag{5.29}$$

Taking into account, the right tank equation of the square in eq. (5.29) we set

$$y_N = \frac{\sqrt{3}}{2}a - \sqrt{3}x, \quad x_E = \frac{a}{2} - \frac{y}{\sqrt{3}}.$$

Recall that here, a is the length of the side of an equilateral triangle. Figure 5.9 shows the graph of the velocity distribution according to formula (5.29) and Fig. 5.10 shows the exact solution for $a = 1$.

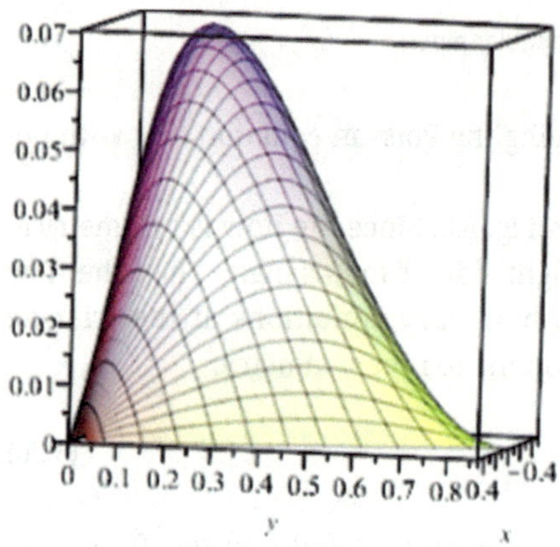

Fig. 5.9: Approximate solution.

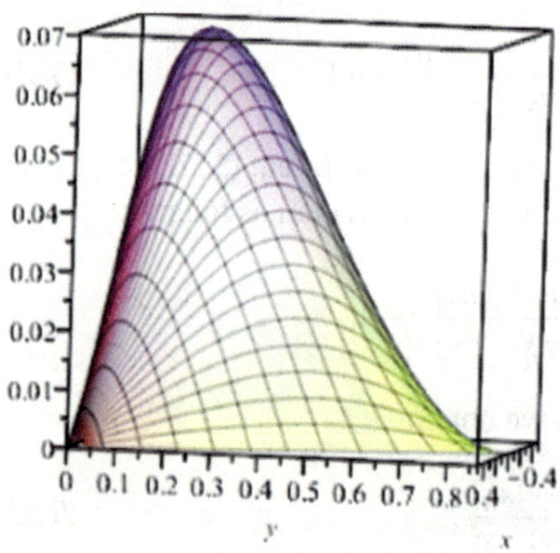

Fig. 5.10: Exact solution.

Using the method of moving nodes, it is possible to obtain solutions to the problem of flow in pipes with different configurations. A simple form of the solution makes it possible to analyze the solutions.

5.2.2 Solution of the open flow problem with a rectangular cross section

Let us consider flows in a long channel. There is no pressure drop along the length of the flow, the transverse pressure drop is static, the same in all sections.

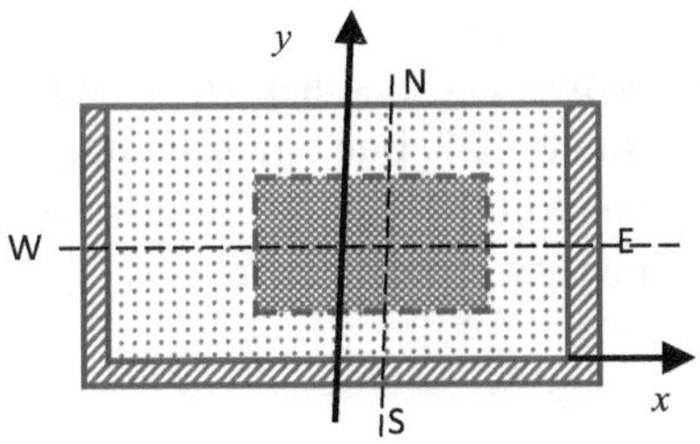

Fig. 5.11: Rectangular channel.

Then from the Navier-Stokes equation [33] we obtain

$$\frac{\partial^2 u}{\partial x^2}+\frac{\partial^2 u}{\partial y^2}=-\frac{\rho g \sin \alpha}{\mu}. \tag{5.30}$$

Let us consider the channel cross section in the form of a rectangle. The channel height is h, and the width is $2b$. No-slip conditions are set on the rigid boundaries. The condition is satisfied on the free boundary:

$$\frac{\partial u}{\partial y}=0. \tag{5.31}$$

Integrating eq. (5.30) over the control volume (Fig. 5.11), we obtain

$$\frac{2}{x_E-x_W}\left[\frac{u_E-u}{x_E-x}-\frac{u-u_W}{x-x_W}\right]+\frac{2}{y_N-y_S}\left[\frac{u_N-u}{y_N-y}-\frac{u-u_W}{y-y_S}\right]=A. \tag{5.32}$$

Here $A=-\frac{\rho g \sin \alpha}{\mu}$.

It follows from the no-slip condition that $u_E=u_W=u_S=0$. To determine u_N, we proceed as follows. In eq. (5.32) we pass to the limit a$y \to y_N$; taking into account $x_E-x_W=2b,\ \ y_N-y_S=h$, we obtain

$$\frac{u_N}{b}\left[\frac{1}{b-x}+\frac{1}{x+b}\right]+\frac{2}{h}\left[\frac{\partial u(x,y_N)}{\partial y}-\frac{u_N}{h}\right]=A.$$

Hence, using eq. (5.31), we determine

$$u_N=-A\frac{h^2(b-x)(x+b)}{2h^2+2(b^2-x^2)}. \tag{5.33}$$

Using the no-slip conditions and eq. (5.33) from eq. (5.32), we obtain

$$u = -A\frac{(b^2 - x^2)y}{y(h-y) + b^2 - x^2}\left[\frac{h-y}{2} + \frac{h(b^2 - x^2)}{2(h^2 + b^2 - x^2)}\right]. \qquad (5.34)$$

5.2.3 Solution of the open flow problem with a trapezoidal cross section

Let us now consider the velocity distribution over a trapezoidal channel. The channel characteristics and the coordinate system are shown in Fig. 5.12. Due to the symmetry of the flow, relative to the axis, we consider the right half of the channel. In this case, using the same technique, we obtain the solution of the flow problem as follows [7]:

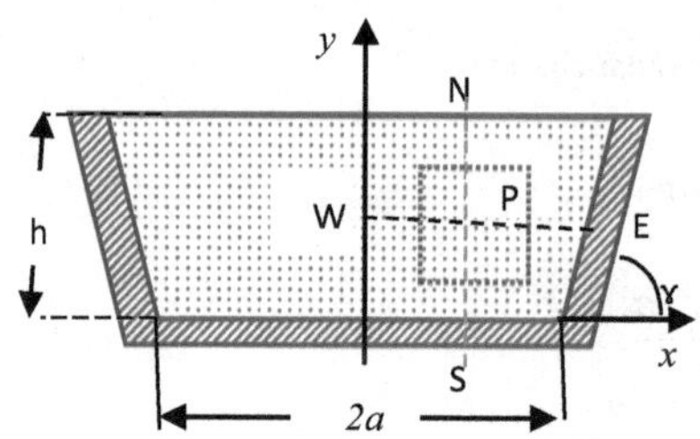

Fig. 5.12: Trapezoidal channel.

$$u = \frac{x_E - x}{x(x_E - x) + (b - y)(y - y_S)}\left[-\frac{A}{2}x(h-y)(y-y_S) + \frac{(h-y)(y-y_S)u_W}{x_E} + xu_N\right]. \qquad (5.35)$$

Here

$$u_N = \frac{A}{2}\left[x_E - x + \frac{(h-y)(y-y_S)x_E^2}{V}\right]\frac{V(b-y_S)(y-y_S)}{x(y-y_S)^3x_E^2 - \left[x(x_E-x) + (y-y_S)^2\right](y-y_S)V}$$

$$u_W = \frac{A}{2}(x_E - x) + \frac{x(x_E - x) + (y-y_S)^2}{x(y-y_S)^2}u_N, \quad V = x_E^2 + (h-y)(y-y_S),$$

$$y_S = \begin{cases} 0 & \text{if} \quad x_E \le a, \\ (x_E - a)tg\gamma & \text{if} \quad x_E > a, \ a < x_E \le a + h/tg\gamma. \end{cases}$$

5.3 Improving the approximate solution for boundary value problems

This section presents an improvement of the approximate solution using the movable node method. In the previous sections, using the movable finite-difference method or the control volume method, an approximate solution was obtained for elliptic boundary value problems. Here, an improvement of the solution is proposed using a partial approximation of differential operators. The main idea is that in a partial differential equation with respect to some variable, we make an approximation in order to reduce the partial differential equation to an ordinary equation.

The approximate solution for problems (5.1) and (5.2) given in eq. (5.6) is simpler and more convenient. This solution is obtained by approximating the partial derivatives in eq. (5.1) with respect to both variables. It is easy to see that this approach can also be applied to multidimensional problems. This section is devoted to improving the approximate solution, which is achieved by partial approximation of the partial derivatives appearing in the differential equation. For example, in the differential eq. (5.1), we approximate the second derivative with respect to y in order to obtain a more improved solution.

5.3.1 Obtaining an approximate solution to the problem with approximation with respect to the variable *x*

To increase the accuracy of the obtained approximate solution, we perform a partial approximation in eq. (5.1). This approach is called the method of lines in numerical analysis [15, 20]. We will combine this approach with the method of moving nodes. In eq. (5.1), we approximate the second derivative with respect to x with a moving node and obtain the following expression (for brevity, $f(x,y) = 0$):

$$\frac{2}{b-a}\left[\frac{u_b(y)-u(x,y)}{b-x}+\frac{u(x,y)-u_a(y)}{x-a}\right]+\frac{d^2u}{dy^2}\approx 0.$$

From here, we obtain an ordinary differential equation when replacing the approximate equality with the exact one:

$$\frac{d^2U_1}{dy^2}-\frac{2}{(b-x)(x-a)}U_1=-\left[\frac{2u_b(y)}{(b-a)(b-x)}+\frac{2u_a(y)}{(b-a)(x-a)}\right]. \tag{5.36}$$

In eq. (5.36), U_1 is an approximate solution. Solving eq. (5.36), we obtain a more precise solution. Here, we assume that y is a variable, and we solve a second-order differential equation with respect to this variable, and we consider x to be a constant.

We introduce the following notations in eq. (5.36):

$$A=\frac{2}{(b-x)(x-a)},\quad B=\frac{2u_b(y)}{(b-a)(b-x)}+\frac{2u_a(y)}{(b-a)(x-a)}.$$

We solve this eq. (5.36) as an ordinary differential equation with respect to the variable y. Taking into account the boundary conditions, the general solution will have the following form:

$$U_1=C_1\exp\left(\sqrt{A}y\right)+C_2\exp\left(-\sqrt{A}y\right)+Q(x,y). \tag{5.37}$$

The particular solution $Q(x,y)$ depends on the assignment of the right-hand side of eq. (5.36):

$$C_2 = \frac{u_d(x) - (u_c(x) - Q(x,c))\exp(\sqrt{A}(d-c)) - Q(x,d)}{\exp(-\sqrt{A}d) - \exp(-\sqrt{A}(d-2c))},$$

$$C_1 = u_c(x)\exp\left(-\sqrt{A}c\right) - C_2\exp\left(-2\sqrt{A}c\right) - Q(x,c)\exp\left(-\sqrt{A}c\right).$$

5.3.2 Obtaining an approximate solution to the problem by approximating the variable *y*

The method for obtaining an approximate solution to the problem is similar to that in the previous paragraph. Now we approximate the second derivative with respect to y, and as a result, we obtain the following solution:

$$U_2 = C_1\exp\left(\sqrt{E}x\right) + C_2\exp\left(-\sqrt{E}x\right) + L(x,y), \tag{5.38}$$

$$C_2 = \frac{u_b(y) - (u_a(y) - L(a,y))\exp(\sqrt{D}(b-a)) - L(b,y)}{\exp(-\sqrt{D}b) - \exp(-\sqrt{D}(b-2a))},$$

$$E = \frac{2}{(d-y)(y-c)}, \quad D = \frac{2u_d(x)}{(d-c)(d-y)} + \frac{2u_c(x)}{(d-c)(y-c)}.$$

5.3.3 Solution of test problems

***Problem* 1.** Consider the heat equation in the region $[0,a] \times [0,b]$

$$\Delta T = 0$$

with boundary conditions:

$$T(0,y) = T_0 + (T_3 - T_0)\frac{y}{b}, \quad T(a,y) = T_1 + (T_2 - T_1)\frac{y}{b},$$

$$T(x,0) = T_0 + (T_1 - T_0)\frac{x}{a}, \quad T(x,b) = T_3 + (T_2 - T_3)\frac{x}{a}.$$

Here

$$T(0,0) = T_0, \ T(a,0) = T_1, \ T(0,b) = T_2, \ T(a,b) = T_3$$

are the temperatures of the corner points.

Exact solution of this problem is

$$T(x,y) = T_0 + \frac{T_1 - T_0}{a}x + \frac{T_3 - T_0}{b}y + \frac{T_0 + T_2 - T_1 - T_3}{ab}xy.$$

We approximate the Laplace equation by a moving difference scheme:

$$\frac{2}{a}\left(\frac{T(a,y) - T(x,y)}{a-x} - \frac{T(x,y) - T(0,y)}{x}\right) + \frac{2}{b}\left(\frac{T(x,b) - T(x,y)}{b-y} - \frac{T(x,y) - T(x,0)}{y}\right) = 0.$$

The solution of this equation coincides with the exact solution. No further improvement is required.

Problem 2. Consider the Poisson equation

$$\Delta u = 6x - 6y,$$

with boundary conditions in the region [0,1] × [0,1]:

$$u(0, y) = -y^3, \quad u(1,y) = 1 - y^3, \quad u(x, 0) = x^3, \quad u(x,1) = x^3 - 1.$$

Exact solution of the problem is $u = x^3 - y^3$.

Using a standard moving node (approximation in both variables), it is easy to obtain an approximate solution:

$$U = \frac{xy(y-x) + y^4(1-y) + x^4(x-1)}{y(y-1) + x(x-1)}. \tag{5.39}$$

Using the approximate solution according to formula (5.37), we obtain

$$U_2 = C_1 \exp\left(\sqrt{A}x\right) + C_2 \exp\left(-\sqrt{A}x\right). + x^3 - y + 3y^2(1-y), \tag{5.40}$$

where

$$C_2 = \frac{(2y^3 - 3y^2 + y)(1 - \exp(\sqrt{A})}{y(1-y)},$$

$$C_1 = 2y^3 - 3y^2 + y - C_2, \quad A = \frac{2}{y(1-y)}.$$

Figure 5.13 shows the exact and approximate solution of problem 2. It is evident from the graph that these solutions are close. Figure 5.13(c) shows the graph of the difference between the exact and standard approximate solution (5.39). The one shown in Fig. 5.13(d) shows an improvement in the approximate solution.

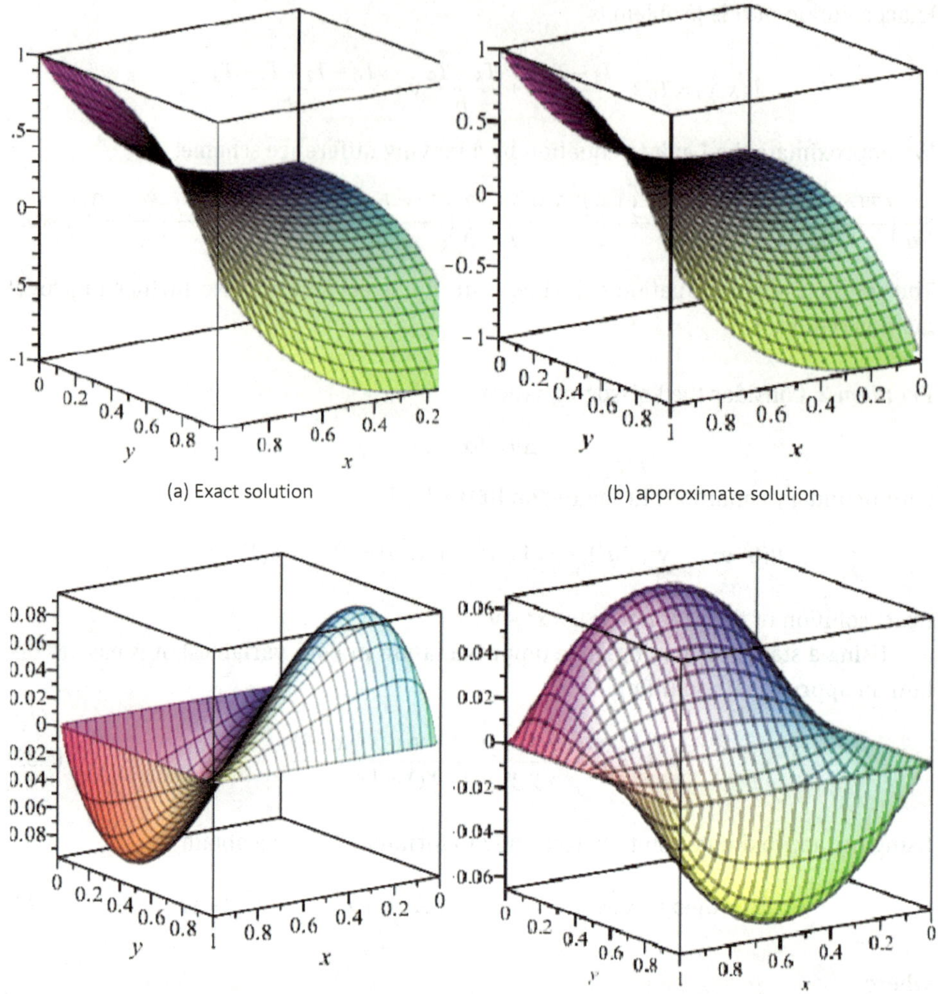

(a) Exact solution

(b) approximate solution

(c) the difference between the exact and approximate solution obtained by formula (5.39)

(d) the difference between the exact and approximate solution obtained by formula (11)(5.40)

Fig. 5.13: Solution of problem (5.2).

Problem 3. Solve the Poisson equation

$$\Delta u = 8\cos^2(x+y) - 4, \tag{5.41}$$

in the domain [0,1] × [0,1] with boundary conditions:

$$\begin{aligned} &u(0, y) = \sin^2 y, \quad u(1,y) = \sin^2(1+y), \\ &u(x, 0) = \sin^2 x, \quad u(x,1) = \sin^2(1+x). \end{aligned} \tag{5.42}$$

Exact solution is

$$u = \sin^2(x+y).$$

The standard approximate solution, obtained on the basis of approximation in both variables, has the form:

$$U = \frac{y(1-y)\left[x\cdot\sin^2(y+1) + (1-x)\sin^2(y)\right]}{M} + \frac{x(1-x)\left[y\cdot\sin^2(x+1) + (1-y)\sin^2(x)\right]}{2M} + \frac{x(1-x)y(1-y)}{M}\left[8\cos^2(x+y) - 4\right], \tag{5.43}$$

where

$$M = x(1-x) + y(1-y).$$

The approximate solution according to formula (5.37) gives

$$U_2 = C_1 \exp\left(\sqrt{A}x\right) + C_2 \exp\left(-\sqrt{A}x\right). + \tilde{u}. \tag{5.44}$$

In eq. (5.44), $A = 2/[y(1-y)]$, and $\tilde{u}$ the particular solution, which is defined as

$$\tilde{u} = a_1\cos(2(1+y)) + a_2\cos(2x) + a_3\cos(2(x+y)) + 1/2,$$

$$a_1 = \frac{-y}{y^2(1-y)+2}, \quad a_2 = \frac{y-1}{y^2(1-y)+2}, \quad a_3 = \frac{-4y(1-y)}{y^2(1-y)+2}.$$

C_1, C_2 are determined based on the boundary conditions.

Figure 5.14 shows the graphs of the solution of problem (5.3), Fig. 5.14(a) corresponds to the exact solution and Fig. 5.14(b) to the approximate solution, obtained on the basis of the standard approximation of derivatives with a moving difference ratio. The closeness of these solutions (the difference between the exact and approximate solution (5.43)) is reflected in Fig. 5.14(c), which shows good agreement of the results (the maximum error in absolute value does not exceed 0.06). If we use the improved approximate solution according to formula (5.44), an increase in the accuracy of the approximate solution is almost two times (Fig. 5.14(d)).

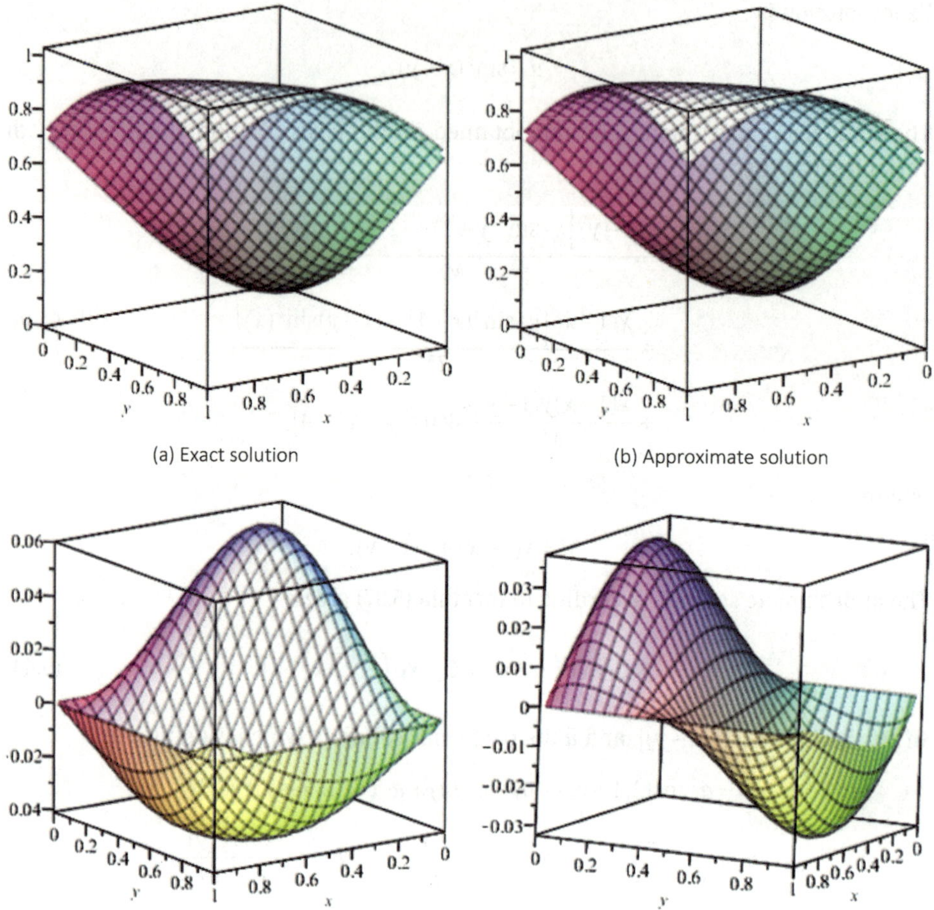

(a) Exact solution

(b) Approximate solution

(c) Difference between exact and approximate solution obtained by formula (5.43)

(d) Difference between exact and approximate solution obtained by formula (5.44)

Fig. 5.14: Solution of problem (5.3).

***Problem* 4.** Solve the Laplace equation

$$\Delta u = 0,$$

in the domain $[0,1] \times [0,\pi/2]$ with boundary conditions:

$$\begin{aligned} u(0, y) &= \cos y, \quad u(1,y) = \cos y/e, \\ u(x, 0) &= \exp(-x), \quad u(x,\pi/2) = 0. \end{aligned} \tag{5.45}$$

Exact solution is

$$u = \exp(-x)\cos y.$$

The standard scheme for moving nodes gives the following result:

$$U = \frac{y(\pi/2-y)[x/e+1-x]\cos y + 2/\pi \cdot x(1-x)[\pi/2-y]\exp(-x)}{x(1-x)+y(\pi/2-y)}. \tag{5.46}$$

The results of calculations using the standard scheme are given by

$$U_2 = C_1\exp\left(\sqrt{A}y\right) + C_2\exp\left(-\sqrt{A}y\right). + B\cos y. \tag{5.47}$$

Here the notation is introduced:

$$A = \frac{2}{x(1-x)}, \quad B = \frac{2(e^{-1}x+1-s)}{-x^2+x+2},$$

$$C_1 = \frac{\exp(-\sqrt{A}\pi/2)(B-\exp(-x))}{\exp(\sqrt{A}\pi/2)-\exp(-\sqrt{A}\pi/2)}, \quad C_2 = \exp(-x) - C_1 - B.$$

Figure 5.15 shows the graphs of the difference between the exact and approximate solutions obtained using formulas (5.46) and (5.47), which show an almost twofold improvement in the solution.

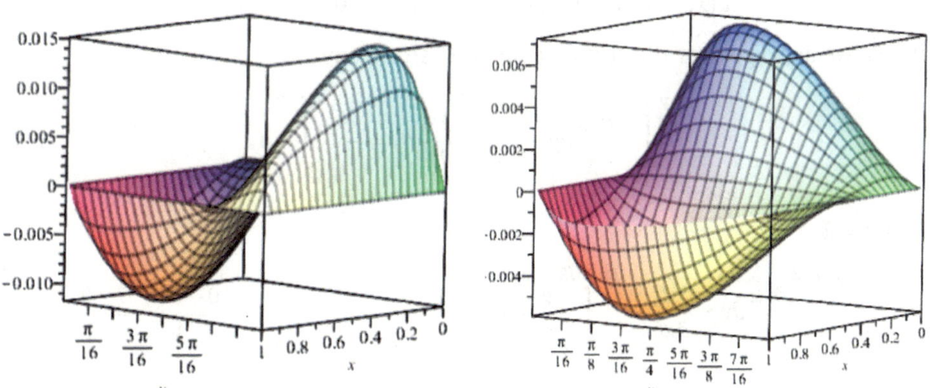

(a) graph of the difference between the exact and standard approximations

(b) graph of the difference between the exact and improved approximations

Fig. 5.15: Graphs of the difference between the exact and approximate solutions.

The calculation of the root-mean-square error using the standard and improved approaches is 0.00581 and 0.00382, respectively. The root-mean-square errors are calculated using the formula:

$$Er = \sqrt{\frac{\sum_{i=1}^{N}\sum_{j=1}^{M}(u_{ij} - U_{ij})^2}{NM}}$$

for $N = 10$ and $M = 10$ and $u_{ij} = u(x_i, y_j)$.

Thus, the partial approximation obtained in the second and third methods gives a better approximation to the exact solution. The main advantage of the first method (the method of moving nodes) is the simplicity of this method, and it can be used in engineering and technical calculations.

5.3.4 Improving the solution to the problem of flow in a rectangular pipe

We improve the solution to the problem considered in Section 5.1.3, the problem of incompressible viscous fluid flow through a rectangular pipe [2].

The problem in dimensionless form is reduced to the following equation (the notation is the same as in Section 5.1.3):

$$\frac{\partial^2 u}{\partial x^2} + \frac{\partial^2 u}{\partial y^2} = -1. \tag{5.48}$$

Boundary conditions for eq. (5.48)

$$u(x, -1) = 0, \quad u(x, 1) = 0, \quad u(-\sigma, y) = 0, \quad u(\sigma, y) = 0. \tag{5.49}$$

If s >1, we average over the cross-section x. Then we have

$$\int_{(y-1)/2}^{(y+1)/2} \frac{\partial^2 u}{\partial x^2} dy + \int_{(y-1)/2}^{(y+1)/2} \frac{\partial^2 u}{\partial y^2} dy = -\int_{(y-1)/2}^{(y+1)/2} dy. \tag{5.50}$$

Assuming,

$$\int_{(y-1)/2}^{(y+1)/2} \frac{\partial^2 u}{\partial x^2} dy \approx \frac{\partial^2 U}{\partial x^2}; \quad \int_{(y-1)/2}^{(y+1)/2} \frac{\partial^2 u}{\partial y^2} dy = \left.\frac{\partial u}{\partial y}\right|_{(y+1)/2} - \left.\frac{\partial u}{\partial y}\right|_{(y-1)/2} \approx \frac{0 - U}{1 - y} - \frac{U - 0}{y + 1}.$$

Here U is the velocity averaged over the cross section. Thus, we obtain the averaged equation in the form:

$$\frac{\partial^2 U}{\partial x^2} - \frac{U}{1-y} - \frac{U}{y+1} = -1. \tag{5.51}$$

In the equation, considering y as a parameter, taking into account the boundary conditions, we obtain

$$U = C_1 \exp\left(-\sqrt{k}\sigma x\right). + C_2 \exp\left(\sqrt{k}\sigma x\right) + k^{-1}. \tag{5.52}$$

Here

$$C_2 = -\left[k\left(\exp\left(2\sqrt{k}\sigma\right) - \exp\left(-2\sqrt{k}\sigma\right)\right)\right]^{-1}\left(\exp\left(\sqrt{k}\sigma\right) - \exp\left(-\sqrt{k}\sigma\right)\right),$$
$$C_1 = -C_2 \exp\left(2\sqrt{k}\sigma\right) - k^{-1}\exp\left(\sqrt{k}\sigma\right), \quad k = 2(1-y^2).$$

The exact solution of the problem has the form:

$$u = \frac{16\sigma^2}{\pi^3}\sum_{n=0}^{\infty}\frac{(-1)^n}{(2n+1)^3}\left[1 - \frac{ch\left(\frac{2n+1}{2}\frac{\pi}{\sigma}y\right)}{ch\left(\frac{2n+1}{2}\frac{\pi}{\sigma}\right)}\right]\cos\left(\frac{2n+1}{2}\frac{\pi}{\sigma}x\right). \tag{5.53}$$

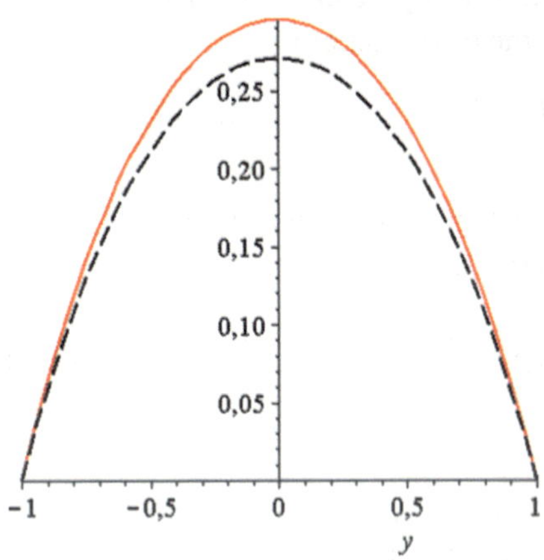

Fig. 5.16: Comparison of the solution on the cross section $x = 0$ according to (1.56).

Fig. 5.16 shows a comparison of (5.52) and (3.53). The maximum absolute difference between the exact and approximate solutions is 0.024. Thus, with partial approximation, the result increases twofold.

5.3.5 Flow at the inlet section of the pipe

With appropriate simplifications, the flow of a viscous incompressible fluid in dimensionless form is described by the following differential equation:

$$u\frac{\partial u}{\partial x} = -\frac{1}{Re}N + \frac{1}{Re}\left(\frac{\partial^2 u}{\partial y^2} + \frac{\partial^2 u}{\partial x^2}\right). \tag{5.54}$$

Here, $N = -12$, is the pressure drop providing a unit flow rate over the pipe cross section and Re is the Reynolds number. The equations are considered in the region $D{:}\ -0{,}5 < x < 0.5,\quad 0 < y < L$ (Fig. 5.17). Boundary conditions for eq. (5.54):

$$u(0,y) = 1;\quad u(L,y) = 1{,}5(1-4y^2);$$
$$u(x,-0{,}5) = 0;\quad u(x,0.5) = 0.$$

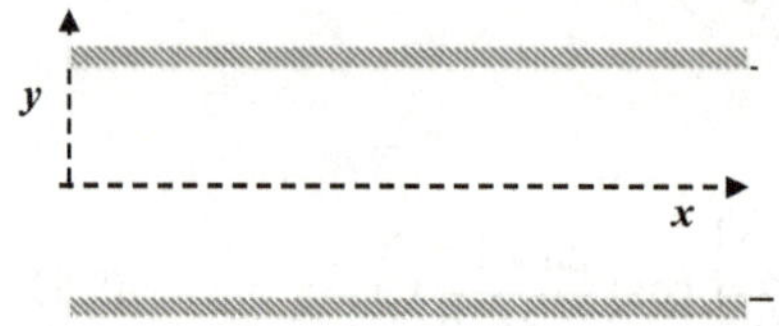

Fig. 5.17: Flow region.

We linearize the convective term as follows (the velocity before the derivative is replaced by the average value of the velocity over the cross section):

$$u\frac{\partial u}{\partial x} \approx \frac{\partial u}{\partial x}.$$

Approximating in eq. (5.54) over the liquid volume $(y+0.5)/2 < y < (y-0.5)/2$, we obtain an ordinary equation, solving which we obtain an approximate solution:

$$u = C_1 \exp(k_1 x) + C_2 \exp(k_2 x) - \frac{1-4y^2}{8}N, \tag{5.55}$$

where

$$k_{1,2} = \frac{Re}{2}\left(1 \pm \sqrt{1 + \frac{32}{Re^2(1-4y^2)}}\right).$$

For comparison, solution (5.54) was made with the numerical method.

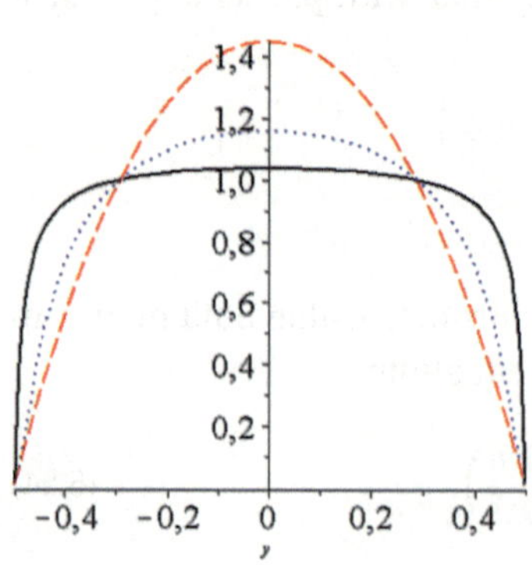

Fig. 5.18: Approximate solution based on eq. (5.55). Velocity profiles corresponding to sections: $x = 0,1;\ 0,5;\ 3$ ($Re = 1$, $L = 5$).

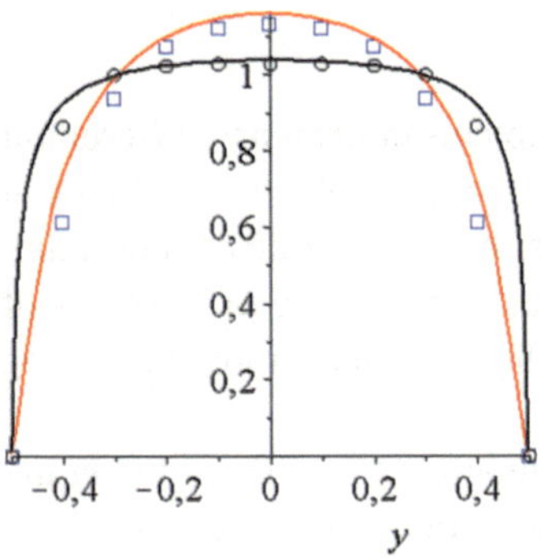

Fig. 5.19: Comparison of approximate and numerical solutions ($Re = 1$, $L = 5$).

Figure 5.18 shows the velocity profiles obtained on the basis of the approximate solution. The solid curve corresponds to the section $x = 0.1$, the dotted curve corresponds to the section $x = 0.5$ and the dashed curve corresponds to the section $x = 3$. Figure 5.19 shows a comparison of the approximate and numerical solutions of equation (5.54). The solid lines correspond to solution (5.55) and the dotted lines to the numerical solution (velocity profiles are given for section $x = 0.1$ and $x = 0.5$). Thus, the approximate solution reflects well the behavior of the velocity profile across the pipe sections.

Conclusions

The proposed method of movable nodes for solving problems of ordinary differential equations (the Cauchy problem and two-point boundary value problems) and partial differential equations (initial-boundary value problems for a parabolic equation and boundary value problems for elitist equations) allows us to assert the universality of this approach. The peculiarity of this method is its simplicity of application for solving problems of differential equations.

The introduction of movable nodes for difference operators of differential equations in combination with numerical methods (the finite-difference method and control volume methods) allowed us to obtain an approximate analytical solution to problems of differential equations.

Solutions to problems with one movable node are a very simple and universal way to obtain a rough analytical representation of the solution to problems of differential equations.

To clarify the solution obtained with one movable node, in the case of boundary value problems of ordinary differential equations, multipoint movable nodes can be successfully used. Multipoint movable nodes allow constructing compact schemes of high order of accuracy.

Using a combination of movable node methods and Richardson extrapolation, compact schemes with high resolution for convective-diffusion problems are constructed.

To improve the accuracy of the obtained approximate solution in the case of problems of partial differential equations, it is convenient to use the method of the method of lines in combination with the method of movable nodes.

Here, the method of movable nodes is mainly used for linear problems. In the future, nonlinear problems of differential equations with movable nodes should be considered. The method of estimating the obtained solutions should also be investigated.

https://doi.org/10.1515/9783111308173-007

References

[1] Anderson D.A., Tannehill J.C., Pletcher R.H. Computational Fluid Mechanics and Heat Transfer. Second Edition, Taylor & Francis.

[2] Umurdin D. Computing Technology of a Method of Control Volume for Obtaining of the Approximate Analytical Solution to One-Dimensional Convection-Diffusion Problems. Open Access Library Journal 5, (2018), 1–6. doi: 10.4236/oalib.1104962.

[3] Dalabaev U. Computing technology of a method of control volume for obtaining of the approximate analytical solution to one-dimensional convection-diffusion problems. Open Access Library Journal (2018), 1104962.

[4] Dalabaev U. Application of the method of moving nodes to the solution of applied boundary problems. Bulletin of the Institute of Mathematics. – Tashkent 6, (2018), 5–9.

[5] Dalabaev U. Difference-analytical method for approximate solution of the Dirichlet problem. Synergy of Sciences 21, (2018), 344–349. SINERGIYA NAUK. URL: http://synergy-journal.ru/archive/article1949.

[6] Dalabaev U. Increasing the accuracy of the difference scheme using the Richardson extrapolation based on the movable node method. Academic Journal of Applied Mathematical Sciences ISSN(e): (2020), 2415–2188, ISSN(p): 2415–5225, doi: 10.32861/ajams.68.204.212.

[7] Dalabaev U., Hasanova D. Engineering method for calculating fluid flow in pipes with different cross sections in energy and technological processes. E3S Web of Conferences 371, (2023), 03021. https://doi.org/10.1051/e3sconf/202337103021.

[8] Darwish M., Asmar D., Moukalled F. A comparative assessment within a multigrid environment of segregated pressure-based algorithms for fluid flow at all speeds. Numerical Heat Transfer, Part B 45, (2003), 49–74.

[9] Dulan E.M.D., Shilders U. Equal Numerical Methods for Solving Boundary Layer Problems. Moscow: Mir Publishing House, (1983), 200.

[10] Faizullaev D.F. Laminar Motion of Multiphase Media in Conduits. Springer US, ISBN 978-1-4899-4832-8, (1969).

[11] Ferreira V.G., de Queiroz R.A.B., Lima G.A.B., Cuenca R.G., Oishi C.M., Azevedo J.L.F., McKee S. A bounded upwinding scheme for computing convection-dominated transport problems. Computers & Fluids 57, (2012), 208–224.

[12] Ferreira V.G., Kurokawa F.A., Queiroz R.A.B., Kaibara M.K., Oishi C.M., Cuminato J.A., Castelo A., Tome M.F., McKee S. Assessment of a high-order finite difference upwind scheme for the simulation of convection–diffusion problems. International Journal for Numerical Methods in Fluids. (2009).

[13] Gao W., Li H., Jian Y. An oscillation-free high order TVD/CBC-based upwind scheme for convection discretization. Numerical Algorithms 59, (2012), 29–50.

[14] Gaskell P.H., Lau A.K.C. Curvature-compensated convective transport: SMART, a new boundedness preserving transport algorithm. International Journal for Numerical Methods in Fluids 8, (1988), 617641.

[15] Hamdi S., Schiesser W.E. †, Griffiths G.W. Method of Lines. Scholarpedia 2(7), (July 9, 2009), 2859.

[16] Hayase T.A., Humphrey J.A.C., Greif R. Consistently formulated QUICK scheme for fast and stable convergence using finite-volume iterative calculation procedure. Journal of Computational Physics 98, (1992).

[17] Leonard B.P. A stable and accurate convective modeling procedure based on quadratic upstream interpolation. Computer Methods in Applied Mechanics and Engineering 19, (1979), 59–98.

[18] Leonard B.P. A survey of Finiye difference with upwind for numerical modeling of the incompressible convective diffusion equation. Taylor C., Morgan K. (Eds.), Computational Techniques in Transient and Turbulent Flows. Swansea, UK: Pineridge Press, (1981), pp. 1–35.

https://doi.org/10.1515/9783111308173-008

[19] Leonard B.P. The ULTIMATE conservative difference scheme applied to unsteady one-dimensional advection. Computer Methods in Applied Mechanics and Engineering 88, (1991), 17–74.

[20] Liskovets O.A. Method of Lines. Differential Equations 1(12), (1965), 1662–1678.

[21] Loitsyansky L.G. Mechanics of Liquid and Gas. Moscow: Drofa, (2003), p. 840.

[22] Marchuk G.I. Methods of Computational Mathematics. Moscow, (1977).

[23] Na T.S. Numerical Methods for Solving Boundary Layer Problems. M.: Mir, (1982). s. 294.

[24] Nigmatulin R.I. Fundamentals of the Mechanics of Heterogeneous Media. Moscow, Izdatel'stvo Nauka. Russian, (1978), 336.

[25] Patankar S. Numerical Heat Transfer and Fluid Flow. CRC Press, Published January 1, by.(1980), ISBN 9780891165224.

[26] Rasulov A., Dalabaev U. Computational technology for improving the quality of difference schemes based on moving nodes. Journal of Physics: Conference Series 1860 (2021), doi: 10.1088/1742-6596/1860/1/012026.

[27] Samarsky A.A. Introduction to the Numerical Methods. Moscow: Nauka, (1982).

[28] Samarsky A.A., Vabishchevich P.N. Numerical Methods for Solving Convection-diffusion Problems. Moscow: Book House "LIBROKOM", (2015), p. 248.

[29] Tikhonov A.N., Samarskiy A.A. Uravneniya Matematicheskoy Fiziki, Uchebnoye Posobiye Dlya Vuzov. 5th edition, stereotip. M.: Nauka, (1977), s. 735, ill.

[30] Umurdin D., Dilfuza X. Comparative analysis of schemes with movable nodes for a parabolic equation. European Journal of Applied Sciences 12(6), (2024), 344–352.

[31] Umurdin D., Malika I. Moving Node Method for Differential Equations. "Numerical Simulation". Dr. Ali Soofastaei Intech Open edited by, Web of Science Reviewed, p. 62, August 23rd, 2022 Published: February 17th, 2023, DOI: 10.5772/intechopen.107340 https://www.intechopen.com/online-first/85510.

[32] Leer B.V. Towards the ultimate conservative difference scheme. II. Monotonicity and conservation combined in a second order scheme. Journal of Computational Physics 14, (1974), 361–370.

[33] Leer B.V. Towards the ultimate conservation difference scheme V. A second order sequel to Godunov's method. Journal of Computational Physics 23, (1977), 101–136.

[34] Zho J.A. Low-diffusive and oscillation free convection scheme. Communications in Applied Numerical Methods 7, (1991), 225–232.

[35] Zhu J., Rodi W. A low-dispersion and bounded convection scheme. Computer Methods in Applied Mechanics and Engineering 92, (1991), 87–96.

Index

https://doi.org/10.1515/9783111308173-009

De Gruyter Series in Applied and Numerical Mathematics

Volume 10
Cholmin Sin, Evgenii S. Baranovskii
Regularity Theory for Generalized Navier–Stokes Equations. Non-Newtonian Fluids with Variable Power-Law, 2025
ISBN 978-3-11-150876-4, e-ISBN (PDF) 978-3-11-150890-0, e-ISBN (EPUB) 978-3-11-150948-8

Volume 9
Habib Ammari, Bryn Davies
Metamaterial Analysis and Design. A Mathematical Treatment of Cochlea-Inspired Sensors, 2023
ISBN 978-3-11-078404-6, e-ISBN (PDF) 978-3-11-078496-1, e-ISBN (EPUB) 978-3-11-078513-5

Volume 8
Zhi-Zhong Sun, Qifeng Zhang, Guang-hua Gao
Finite Difference Methods for Nonlinear Evolution Equations, 2023
ISBN 978-3-11-079585-1, e-ISBN (PDF) 978-3-11-079601-8, e-ISBN (EPUB) 978-3-11-079611-7

Volume 7
Roland Glowinski, Tsorng-Whay Pan
Numerical Simulation of Incompressible Viscous Flow. Methods and Applications, 2022
ISBN 978-3-11-078491-6, e-ISBN (PDF) 978-3-11-078501-2, e-ISBN (EPUB) 978-3-11-078505-0

Volume 6
Bruno Després
Neural Networks and Numerical Analysis, 2022
ISBN 978-3-11-078312-4, e-ISBN (PDF) 978-3-11-078318-6, e-ISBN (EPUB) 978-3-11-078326-1

Volume 5/1
Alexander V. Bobylev
Kinetic Equations. Volume 1: Boltzmann Equation, Maxwell Models, and Hydrodynamics beyond Navier–Stokes, 2020
ISBN 978-3-11-055012-2, e-ISBN (PDF) 978-3-11-055098-6, e-ISBN (EPUB) 978-3-11-055017-7

Volume 5/2
Alexander V. Bobylev
Kinetic Equations. Volume 2: Landau Equation, Boltzmann-Type Equations, Discrete Models, and Numerical Methods, 2024
ISBN 978-3-11-055013-9, e-ISBN (PDF) 978-3-11-055100-6, e-ISBN (EPUB) 978-3-11-055014-6

Volume 4
Claude Le Bris, Pierre-Louis Lions
Parabolic Equations with Irregular Data and Related Issues. Applications to Stochastic Differential Equations, 2019
ISBN 978-3-11-063313-9, e-ISBN (PDF) 978-3-11-063550-8, e-ISBN (EPUB) 978-3-11-063314-6

Volume 3
Dominic Breit, Eduard Feireisl, Martina Hofmanová
Stochastically Forced Compressible Fluid Flows, 2018
ISBN 978-3-11-049050-3, e-ISBN (PDF) 978-3-11-049255-2, e-ISBN (EPUB) 978-3-11-049076-3

